Politics and the Theory of Spontaneous Order

The theory of spontaneous order conceptualises and explains a number of institutional and social phenomena that are not an intended effect of either individual decisions or a collective consensus but an unplanned outcome of interactions between people pursuing their own aims. Drawing on these insights, this book demonstrates the utility of the theory of spontaneous order in explaining many phenomena in political economy and political science.

The book opens with a discussion of the history and development of the theory of spontaneous order, particularly in economics and the Austrian School. The epistemological premises of the theory are then explored including the formulation of the central idea of social individualism. Demonstrating the potential applications of the theory of spontaneous order to politics, core ideas are examined including democracy, fragile states and the concept of the veil of ignorance. Finally, the limitations and constraints of the theory of spontaneous order are also reviewed and discussed.

This book marks a valuable contribution to the literature on political economy, political science, public choice and political philosophy.

Piotr Szafruga holds a PhD in political science from the Institute of Political Science and International Relations of the Jagiellonian University, Poland.

Routledge Frontiers of Political Economy

Power and Influence of Economists
Contributions to the Social Studies of Economics
Edited by Jens Maesse, Stephan Pühringer, Thierry Rossier and Pierre Benz

Rent-Seeking and Human Capital
How the Hunt for Rents is Changing Our Economic and Political Landscape
Kurt von Seekamm Jr.

The Political Economy of State Intervention
Conserving Capital over the West's Long Depression
Gavin Poynter

Intangible Flow Theory in Economics
Human Participation in Economic and Societal Production
Tiago Cardao-Pito

Foundations of Post-Schumpeterian Economics
Innovation, Institutions and Finance
Beniamino Callegari

Distributive Justice and Taxation
Jørgen Pedersen

The China–US Trade War and South Asian Economies
Edited by Rahul Nath Choudhury

Politics and the Theory of Spontaneous Order
Piotr Szafruga

Preventing the Next Financial Crisis
Victor A. Beker

For more information about this series, please visit: www.routledge.com/books/series/SE0345

Politics and the Theory of Spontaneous Order

Piotr Szafruga

LONDON AND NEW YORK

First published 2021
by Routledge
2 Park Square, Milton Park, Abingdon, Oxon OX14 4RN

and by Routledge
52 Vanderbilt Avenue, New York, NY 10017

Routledge is an imprint of the Taylor & Francis Group, an informa business

British Library Cataloguing-in-Publication Data
A catalogue record for this book is available from the British Library

Library of Congress Cataloging-in-Publication Data
A catalog record has been requested for this book

ISBN: 978-0-367-89710-9 (hbk)
ISBN: 978-0-367-72163-3 (pbk)
ISBN: 978-1-003-02067-7 (ebk)

Typeset in Bembo
by Newgen Publishing UK

Contents

Acknowledgements

The book has been published in 2018 in Polish as a monograph. It is based on the doctoral dissertation thesis defended at Institute of Political Science and International Relations of the Jagiellonian University in Kraków.

I would like to thank my family, my supervisor Professor Barbara Krauz-Mozer, Dr. Iwona Jakimowicz-Pisarska, Professor Tadeusz Klementewicz, Professor Witold Kwaśnicki and all the participants of the seminar at the Department of Political Theory and Government Studies of the Jagiellonian University.

Introduction

The concept of spontaneous order refers to social phenomena that are characterised by a form of order and at the same time are the unintended effects of interactions between individuals pursuing their separate goals. Its origins can be traced back to the works of ancient Roman jurists or reflections of medieval scholastics. However, regardless of the chronology, the greatest contribution to the development of this idea seems to have been made by economics. Even before Adam Smith coined the famous phrase about an "invisible hand" guiding human activities in the marketplace, several thinkers saw the existence of an unplanned order in the economic sphere. As Adam Ferguson wrote, such structures are "the result of human action, but not the execution of any human design" (Ferguson 1966, 122). Economists point to the common benefits brought about by these phenomena. The egoistic pursuit of personal benefit and profit by market participants contributes not only to the satisfaction of consumers' needs and mutual coordination of activities, but often promotes the social interest more than a situation in which people deliberately seek to serve the common good.

This approach was used in the formulation of the theory of spontaneous order, making it possible to explain the existence of a number of institutions not resulting from individual decisions or collective consensus aimed at establishing them. It is worth noting that the very existence of such unplanned phenomena is necessarily based on the fact that humans have limited knowledge about the effects of their own actions. Consequently, one subject of this theory is the issue of limitations in the domain of the deliberate creation of social structures.

A noticeable feature of the concept of spontaneous order is that the specific spheres of human activity to which it might refer are not indicated; it includes all forms of human activity in which the issue of unintended effects arises. Thanks to such a perspective the theory of spontaneous order transcends the traditional boundaries between individual disciplines of social science, such as economics, law or political science. But in the light of the above, reflection on unplanned order should be present in all areas of research into social reality. However, this is not the case. The concept of spontaneous order is not widely recognised, and it is used mainly in the field of economics and, to a lesser extent, in the legal sciences. In particular, there is relatively little interest in this idea among political thinkers.

This book is an attempt to address the lack of a broader reflection on unplanned order. It presents possibilities of applying the theory of spontaneous order in describing and explaining political developments and processes. It also presents the criticism of this theory and its limitations in the area of political economy and political science. The subject seems important because, as far as I know, there are currently no publications presenting a systematic and comprehensive approach to this issue.

The main part of the book consists of four chapters.

Part I is an introduction to the subject under discussion. It places the theory of spontaneous order in the context of considerations about the nature of social phenomena. The first chapter presents a reflection on the links between the theory of spontaneous order, associated mainly with economics, and particular disciplines of social sciences, including political science. The second chapter contains a brief historical outline of the development of the idea of spontaneous order, with particular emphasis on the influence of the Austrian School of Economics.

Part II of the book presents the theory of spontaneous order. Chapter 3 focuses on its epistemological assumptions. It discusses the category of human action, which constitutes the praxeological basis of the theory, and the issue of methodological individualism and subjectivism along with an attempt to provide a synthesis of social individualism, an approach formed within the framework of this theory. Further on, the concepts of knowledge and entrepreneurship, and the praxeological pillars of the theory of spontaneous order are discussed along with criticism of the science of human action as the foundation of social sciences, including political science. In particular, the scientific status of the praxeological approach is also discussed. Chapter 4 presents claims, formulated within the framework of the theory, concerning the formation of social order and the limitations placed on its deliberate formation.

Part III contains reflections on the application of the theory of spontaneous order in the study of politics. The starting point is the question of the criteria of what belongs to the pluralistic and multi-paradigmatic area of political science. In particular, the issue of the legitimacy of the proposed research perspective and its conformity with a possible epistemic reference system in these sciences is addressed. Chapter 6 focuses on the issue of the implementation of this theory in research conducted within the framework of public choice theory. This leads to criticism of the concepts of rational ignorance and rational irrationality from the latter theory, defining the mechanisms of choice. The book indicates the possibility of using the theory of spontaneous order as an alternative research approach, one not requiring the assumption of omniscience or irrationality on the part of individuals. Chapter 7 presents the application of the theory of spontaneous order in analysing the effectiveness of mono- and polycentric systems in the coordination of individuals' actions and the formation of social order. It is juxtaposed with similar considerations carried out within public choice theory.

The chapter concludes with a reflection on the use of the theory in relation to the concept of the veil of ignorance. Chapter 8 discusses the application of the theory of spontaneous order to the study of fragile states. Rejecting the state-centric approach, the theory indicates that the stability of social, political, legal and economic structures depends on the compatibility of the state apparatus with endogenous institutions. It provides a broad perspective on the problem of fragile states, embracing the issue of the legitimacy of the state and the consolidation of decision-making processes.

Part IV presents the issue of limitations and obstacles in the application of the theory of spontaneous order to the study of political phenomena. Chapter 9 addresses the issues of using the theory in political science. The chapter also examines the problem of the categorisation of orders as spontaneous or planned in describing complex social reality. It indicates that this problem has its source in the limits of human cognition. Further on, this issue is developed in Chapter 10 that discusses the issue of the structure and functioning of the human cognitive system and the resulting problem of cognitive boundaries.

Part I

The idea of spontaneous order in the history of the social sciences

1 The idea of spontaneous order in reflection on the nature of the social world

The theory of spontaneous order focuses on social phenomena that arise unintentionally as a result of the interactions between people pursuing their separate goals. This means that people, even though they do not act with the intention of creating a given outcome, unintentionally cause its occurrence in the course of their actions. The structures created as a result of such a process constitute a spontaneous order, as included in the name of this theory. The institutions of money, language and customary laws can be listed as examples.[1] Each of them was created in an unplanned way, in the course of interactions serving to achieve the separate goals of individuals.

The existence of unplanned outcomes to which the theory of spontaneous order refers is closely related to the existence of the cognitive limitations of the human mind, which constrain people's freedom to deliberately create various institutions. Consequently, many institutions with a complex structure, which might seem to be a deliberate product of some omniscient decision-making body, actually result from the bottom-up actions of individuals. They cannot be created in a planned way because of their complex structure, exceeding the capabilities of any person or group (Hayek 1988; Huerta de Soto 2010c).

An extremely concise and at the same time clear description of the idea of spontaneous order, which seems to grasp its core meaning, is that proposed by Friedrich Hayek, who defined it as "the results of human action, but not the execution of human design" (Hayek 1967, 96).[2] A noticeable feature of spontaneous order presented in this way is the absence of any indication of specific forms or spheres of human activity to which it might refer. It remains in a general relation to human action, conceived as deliberate behaviour aimed at satisfying the needs of man as such. Thus, the theory of spontaneous order embraces all forms of human activity that can be said to have unintended effects. In fact, this means that it applies to all social phenomena, including political ones.[3]

Despite the broad spectrum of possible areas of research that emerges from the above, the theory of spontaneous order is not widely recognised in any area of social sciences. It is important, therefore, to point out the special relationship between reflection on spontaneous order and the economic and legal sciences. In the context of economics, two fundamental aspects of this relationship can be distinguished. The first is the relationship between this reflection and the so-called Austrian School of Economics. The second is the fact that reflection on spontaneous order has been with regard to mostly market processes, and to a slightly lesser extent jurisprudence.[4]

The importance of the Austrian School of Economics in the context of this book stems from its influence on the development of spontaneous order theory. It should be noted that this issue is by no means redundant with the second aspect from the previous paragraph. The importance of the ideas of the Austrian School goes far beyond the question of examining spontaneous order in economic processes. The Austrian School effectively created the theory of spontaneous order, which resulted directly from its assumptions. In this sense, the theory is an expression of the perception of social reality of the Austrians.[5] At the same time, it indicates that the presentation of the concepts promoted by the Austrian School is important for our further considerations.

The Austrian School represents a heterodox approach to economics. It is assumed to have started with the publication of Carl Menger's work entitled *Principles of Economics* in 1871.[6] The school describes economic processes as dynamic, rejecting static models that interpret the economy in terms of equilibrium and criticising the mathematics of economics. The fundamental issues underlying such an insight into economic and, more generally, social reality are methodological individualism and subjectivism.

For the Austrian School, it is the individual and his or her decisions that are the crucial aspect in the study of the social world. Its representatives point to human choices as the starting point for the study of social reality. This methodological individualism stems from the general perception of economics as a form of learning about human behaviour. It is based on the claim that the only entity that acts and thus makes choices is man. This means that all social phenomena are necessarily the result of the activity of individuals and thus can only be explained through the actions of individual people. This approach, however, does not mean any supremacy of individual over collective phenomena, but only emphasises a form of explaining social reality.

The role of human actions in the perception of social reality makes it crucial to understand the beliefs and preferences behind these actions. For the Austrians, this means that in social sciences facts are the meanings that individuals give to their actions and the surrounding reality, and that the explanation of the social world itself is based on the subjective mental states of individual people. This approach, constituting the subjectivism of the Austrian School, stems from an attempt to create an economics based on the actual human being, seen as a creative actor and driving force of all social processes. This means that social reality is the product of human activity. Thus, an approach proposing that there are objective processes external to man is wrong.[7] Without reference to the meanings given to human activities, understanding of the social world is inadequate. Hence, subjectivism is the key assumption of the Austrian School, and not only in economics (Mises 2008, 20–21).[8] As Ludwig von Mises (2008, 51) pointed out, the aim of the social sciences is to understand the meaning and role of human action. An excellent example of such an approach is the value theory developed by the Austrian School, according to which the value of market goods is subjective and determined by the individuals valuing them. Goods do not have an objective value that goes beyond the subjective estimations of individual market participants.

Another important assumption in the Austrian approach to social reality is the lack of access to full and reliable knowledge on the basis of which individuals take action. This is due to the cognitive limitations of the human mind and also the subjective nature of knowledge possessed and produced by other actors, to which an individual does not have direct access. As a result, the knowledge needed for individuals to take action is incomplete and dispersed among all the actors. This observation was used by representatives of the Austrian School to demonstrate the impossibility of economic calculation in socialism (Hayek 1963; Huerta de Soto 2010; Lavoie 2015). As Mises (1990a) explained, in a system based on public ownership of the means of production, it becomes impossible to determine the value of capital goods according to supply and demand of resources and products. This deprives the central agency of information on the basis of which it could make rational decisions regarding allocation of resources. Bruno Leoni (1961, 90) noted that this claim represented a specific case of a more general phenomenon. In the context of the legal system and opposing the centralised lawmaking mechanism, he pointed out that the problem of the lack of knowledge needed to make decisions was a general feature of structures based on the presence of a central management body.

Subjectivity is a key feature distinguishing the approach of the Austrian School of Economics, and thus also the theory of spontaneous order, to the study of the social world. While the perception of aggregated structures as resulting from the actions of individuals represents methodological individualism, the subjective nature of knowledge and the related creativity of man make him a causal factor. At the same time, emphasising the uncertainty and dispersion of knowledge between individuals, this theory negates human omnipotence in the design of social reality. Thus, it allows for the possibility of unplanned, spontaneous structures arising.

The concept of human action underlying the approach of the Austrian School influences its perception of the role of both economics and all other sciences dealing with the social world. In the case of economics, the subject of research is these human actions, for which a means to achieve the goal takes the form of the process of an exchange of goods (Rothbard 2009, 162). So studying economic processes is not so much a separate discipline of social sciences with its own subject matter, but one of the spheres of reflection on human action, focusing on its course in the sphere of voluntary exchange. This leads to the rejection of the neo-classical approach, according to which this science studies human behaviour conceived as "a relationship between ends and scarce means which have alternative uses" (Robbins 1932, 16).[9] Such an understanding of the subject of research presupposes full knowledge of ends and means, thus reducing the problem of economics to the technical issue of resource allocation, and reducing man to an automaton that responds passively to events

(Huerta de Soto 2008, 5). For the Austrian School, since the individual acts on the basis of subjective beliefs, the aim of economics is to explain how people with different expectations and knowledge are able to coordinate their actions despite these differences. This question is a particular case of the general issue of humans' ability to coordinate, understood as the ability to create order. This means that for the Austrians this issue is not limited to economics. The problem that this theory addresses is crucial for them in terms of the social aspect of man. According to Menger, "perhaps the most noteworthy problem of the social sciences" is "*how can it be that institutions which serve the common welfare and are extremely significant for its development come into being without a common will directed toward establishing them?*" [italics in the original] (Menger 1985, 146). The subjective and dispersed nature of knowledge means that the achievement of the intended ends, as well as coordination of behaviours in all spheres of human activity, takes place under conditions of lack of full information and thus in a situation of uncertainty. Since this concerns all areas of human activity, spontaneous order functions everywhere.

While in the light of the above claim by the Austrian economist unplanned order plays a key role in all areas of social science research, the actual situation is somewhat different. Perhaps the best-known term referring to spontaneous order is the famous phrase about an invisible hand guiding the market coined by Adam Smith.[10] As the author of *The Wealth of Nations* wrote, this hand leads a man seeking his own benefit "naturally, or rather necessarily leads him to prefer that employment [of capital] which is most advantageous to the society" (Smith 1979, 454). Moreover, "[b]y pursuing his own interest he [man] frequently promotes that of the society more effectually than when he really intends to promote it" (Smith 1979, 456). This message is often perceived and criticised as praising a lack of market regulations (Basu 2010; LeRoy 2010). Smith himself pointed out, however, that in order to manage the market, the invisible hand requires the existence of a specific legal framework providing a safeguard against the negative effects of selfish human actions. But regardless of the issue of both Smith's vision of the market and of whether the criticism of contemporary economic mechanisms is right, the words about an invisible hand indicate an important feature characteristic of reflection on spontaneous order, namely its close relation to economics. The development of ideas on the unplanned formation of order to a large extent took place in the context of economic processes. Importantly, this resulted not only from the connection with the Austrian School of Economics and (as its name suggests) its focus on studying economic processes. The characteristic link between the idea of spontaneous order and economics goes far beyond the approach developed by this school. In particular, representatives of the so-called Scottish Enlightenment, of which Smith was a representative, paid much attention to the issue of unintended market effects of individual actions. French researchers, led by Richard Cantillon (1959), also made significant contributions. Previously, however, Spanish scholars from the 16th and 17th centuries, such as Luis Saravia de la Calle (1949) and Juan de Mariana (1768), had played an important role in developing the idea of spontaneous order. These thinkers, based at the University of Salamanca, noted the key role of subjective evaluation in economic processes. The second important area in which the question of unplanned order emerged was jurisprudence. Ancient Roman jurists, such as Cicero (2017), perceived law as resulting from a centuries-long process, representing the knowledge of many generations and going beyond the cognitive abilities of man. In later times, a similar approach can be found mainly in the theories of Anglo-Saxon law based on precedent (common law).

It should be noted that the relationship between spontaneous order and economics goes beyond the connection between reflections on the unplanned formation of institutions and reflections on the nature of economic processes, observed in various periods of history. The scarcity of the latter has often resulted in a failure to notice, or ignorance of, the existence of spontaneous processes. The history of human reflection on society indicates that tendencies to perceive the social world in a static manner and the belief in the possibility of deliberately shaping society usually go hand in hand with a poor development of economic thought or even ignorance of economic processes. An extreme example of such an approach is ancient Greece, and in particular, the legacy of the greatest philosophers of that period, such as Plato and Aristotle.[11]

The existence of a link between reflection on unplanned order and economic and, to a lesser extent, juridical thinking contrasts with the general nature of the idea of spontaneous order as relating to all forms of human activity. With regard to the subject matter of this book, there is relatively little interest in the idea of spontaneous order in the area of political thought.[12] Thus, it seems desirable and advisable to present the history of the formation of the concept of spontaneous order in the context of the general development of social ideas. This approach, forming the content of the next subchapter, will familiarise us with the context of the formation of this concept and thus the reasons for this contrast.

One of the factors explaining the preeminent role of economics in the study of spontaneous processes seems to be the easily perceptible role of the individual in the economic world. The market clearly shows how, without a plan imposed from above, people are able to create a form of order based on the exchange of goods and services enabling them to pursue their individual goals. Thus, the market is an expression of preferences and, consequently, of the will of individual people and the effects of realising this will. This individualistic feature of the market was connected with the influence of market mechanisms on the increase of people's welfare. In particular, the noticeable benefits of trade provoked thinkers to reflect on the matter and explain these mechanisms. While actions of individuals in the market were perceived as self-interested, the increase in wealth demonstrated the mutual benefit of market exchange of goods and services. The connection between the two led to the rejection of the view that the egoistic behaviour of individuals goes against common good, which led to reflections on spontaneous order.

A particular manifestation of "order" created in the course of market processes is the institution of money, resulting from exchange-based interactions between market participants. The concept of price associated with money is important because, unlike many other intersubjective expressions of individual preferences, it is characterised by a linear order. This means that any two elements to which prices are assigned can be ranked and compared. Money is a kind of a bridge making it possible to express a subjective evaluation in the form of an intersubjectively communicable numerical value.

It is worth noting at this point that the amazing regularity of price, which is, after all, based on the subjective evaluation of individuals, made it possible to present market developments by means of numbers. This led economists to propose that market knowledge was objective and that economic sciences could be mathematised. This trend was reinforced by "physics envy", as Philip Mirowski (1991) called it, manifesting itself in the pursuit of accuracy and predictability that exists in the natural sciences, and particularly in physics.[13] Thus, paradoxically, the belief in the possibility of modelling and controlling the social world manifested itself to the greatest extent in a discipline that dominated the development of the spontaneous order theory.

It should be stressed that the development of the idea of spontaneous order is strongly connected with research on economic processes. This is indicated by the sheer amount of space devoted to this issue in this part of the book. However, it is incorrect to say that reflection on the unplanned formation of social institutions is in some fundamental way subordinated or ascribed to economics. In particular, this is determined by the general and universal nature of human activity. The emphasis that has been placed on describing the relationship between reflection on spontaneous order and economics is in part due to the desire to clearly emphasise the lack of such subordination. Rather, the relationship in question indicates a certain interaction in which each of the two elements influences the development of the other.

It seems wrong to attach the theory of spontaneous order to economics or law, but also to any other scientific discipline. As the representatives of the Austrian School of Economics, themselves classified as economists, point out, the study of spontaneous processes is an essential subject of research for social sciences as a whole and cannot be limited to the economic sphere. So the theory goes "across" the contemporary division of social sciences. As Norman Barry pointed out, this division "is itself an important barrier to the acceptance of the doctrine of spontaneous evolution precisely because this theory straddles so many of the artificial boundaries between academic disciplines" (Barry 1982, 8). This observation gives additional significance to the fundamental purpose of the next chapter, namely to present the history of reflection on the idea of spontaneous order. As a natural introduction to the following parts of the book, this part not only shows the development of the concepts underlying this theory, but also facilitates its understanding by placing it in the context of the general development of reflection on the nature of the social world.

Notes

1 A detailed presentation of the conception of spontaneous order can be found in Part II.
2 Hayek borrowed this phrase from Adam Ferguson who in his work *An Essay on the History of Civil Society* used the phrase "the result of human action, but not the execution of any human design" (Ferguson 1966, 122). For more about Adam Ferguson, see Section 3 "The Enlightenment" of Chapter 2.
3 The question of founding the spontaneous order theory on a concept of human action and a praxeological approach is discussed in Chapter 3.
4 The history of the theory of spontaneous order is described in Chapter 2.
5 Unless a different meaning is apparent from the text, the term "the Austrians" or "Austrian" refers to the Austrian School of Economics and its representatives, who themselves are often not of Austrian nationality. The term "Austrian School" is synonymous with the Austrian School of Economics.
6 *Grundsätze der Volkswirtschaftslehre* (Menger 1871).
7 The issue of the cognitive status of the social world is presented in Chapter 3.
8 In particular, Hayek pointed to the key role of subjectivity in the further development of this economics. As he wrote, "It is probably no exaggeration to say that every important advance in economic theory during the last hundred years was a further step in the consistent application of subjectivism" (Hayek 1955, 31).
9 According to Roger Backhouse and Steven Medema (2009, 225), the above approach proposed by Lionel Robbins is one of the most widely accepted definitions of economics.
10 The hand as a metaphor for market mechanisms is mentioned by another 18th-century Italian thinker, Ferdinando Galiani (1728–1787), who in his 1751 work "On Money" ("Della moneta") wrote about the "superior hand" governing economic processes (Galiani 1915).
11 This issue is analysed in Section "Antiquity" of Chapter 2.
12 The issue of the application of the idea of spontaneous order in political science is presented in Part III, while Chapter 9 discusses the limitations related to this application.
13 Mirowski (1991) points out that the development of neoclassical economics was strongly inspired by concepts worked out within physics, especially the principle of conservation of energy.

Bibliography

Backhouse, Roger E., and Steven G. Medema. 2009. Retrospectives: on the definition of economics. *Journal of Economic Perspectives* 23 (1): 221–233. doi: 10.1257/jep.23.1.221.

Barry, Norman P. 1982. The tradition of spontaneous order, literature of liberty. *A Review of Contemporary Liberal Thought* 5 (2): 7–58.

Basu, Kaushik. 2010. *Beyond the Invisible Hand: Groundwork for a New Economics*. Princeton: Princeton University Press.

Cantillon, Richard. 1959. *Essai sur la nature du commerce en general*. London: Frank Cass.

Cicero, Marcus Tullius. 2017. *On the Commonwealth and On the Laws*. 2nd ed. Translated and edited by James Zetzel. Cambridge: Cambridge University Press.

Ferguson, Adam. 1966. *An Essay on the History of Civil Society*. Edinburgh: Edinburgh University Press.

Galiani, Ferdinando. 1915. *Della Moneta*. Bari: Gius Laterza & Figli.

Hayek, Friedrich A. 1955. *The Counter Revolution of Science: Studies on the Abuse of Reason*. New York: The Free Press of Glencoe.

Hayek, Friedrich A. 1963. *Collectivist Economic Planning*. London: Routledge and Kegan Paul.

Hayek, Friedrich A. 1967. *Studies in Philosophy, Politics and Economics*. London: Routledge and Kegan Paul.

Hayek, Friedrich A. 1988. *The Fatal Conceit: The Errors of Socialism*. London: Routledge.

Huerta de Soto, Jesus. 2008. *The Austrian School: Market Order and Entrepreneurial Creativity*. Cheltenham: Edward Elgar Publishing.

Huerta de Soto, Jesus. 2010c. *Socialism, Economic Calculation and Entrepreneurship*. Cheltenham: Edward Elgar Publishing.

Lavoie, Don. 2015. *Rivalry and Central Planning: The Socialist Calculation Debate Reconsidered*. Arlington: Mercatus Center.

Leoni, Bruno. 1961. *Freedom and the Law*. Princeton: David Van Nostrand.

LeRoy, Stephen. 2010. Is the "Invisible Hand" Still Relevant? *Federal Reserve Bank of San Francisco*. www.frbsf.org/economic-research/publications/economic-letter/2010/may/invisible-hand-relevance.

Mariana, Juan. de. 1768. *Discurso sobre las enfermedades de la Compañia*. Madrid: D. Gabriel Ramirez.

Menger, Carl. 1871. *Grundsätze der Volkswirtschaftslehre*. Wien: Wilhelm Braumüller.

Menger, Carl. 1985. *Investigations into the Method of the Social Sciences with Special Reference to Economics*. New York: New York University Press.

Mirowski, Philip. 1991. *More Heat than Light: Economics as Social Physics, Physics as Nature's Economics*. Cambridge: Cambridge University Press.

Mises, Ludwig von. 1990a. *Economic Calculation in the Socialist Commonwealth*. Auburn: Ludwig von Mises Institute.

Mises, Ludwig von. 2008. *Human Action: A Treatise on Economics*. Auburn: Ludwig von Mises Institute.

Robbins, Lionel. 1932. *An Essay on the Nature and Significance of Economic Science*. London: Macmillan.

Rothbard, Murray N. 2009. *Man, Economy, and State with Power and Market*. Auburn: Ludwig von Mises Institute.

Saravia de la Calle, Luis. 1949. *Instrucción de mercaderes*. Madrid: Rvdo. Padre Félix Garcia.

Smith, Adam. 1979. *An Inquiry into the Nature and Causes of the Wealth of Nations*, Vol. 1 of *The Glasgow edition of Works and Correspondence of Adam Smith*, edited by R. H. Campbell, A. S. Skinner, and W. B. Todd. Oxford: Clarendon Press.

2 History of the idea of spontaneous order

The history of the idea of spontaneous order is of particular importance for the topic of this book, because the theory of spontaneous order has not yet found any significant application in political sciences. Moreover, it does not seem to be widely recognised by representatives of these sciences. The main area covered by this theory is economics. This means that the role of historical description goes beyond familiarising the reader with the development of the concepts underlying the theory. Placed in the context of the general pursuit of ideas on the nature of the social world, this description is also helpful in reflecting on the reasons for the existing links between the concept of spontaneous order and particular disciplines of the social sciences, including political science. Thus, it contributes to a better understanding of the theory itself and its application in various areas of research.

Antiquity

It is customary to start reflecting on the history of an idea or issue by reaching back to the ancient Greek period. Undoubtedly, the thinkers of the Hellenic world had a significant impact on the intellectual development of humanity. However, while their influence on the development of epistemology, logic or ethics cannot be overestimated, this cannot be said of their attempts to understand social reality. These attempts, undertaken by many Greek philosophers, were largely unsuccessful. This resulted to a great extent from an important distinction made by the Greeks in the sphere of the capabilities of human cognition. The Greeks are credited with being the first to explain phenomena by pointing not to the arbitrary will of gods, but to a reason. This approach was based on the assumption of the existence of necessary laws governing these phenomena and being cognisable by man. This made it possible to de-mythologise thinking about the world and marked the beginning of science. Importantly, the assumption of the existence of such laws regarded only the natural world, excluding man and the effects of his actions.

One of the elements of philosophical reflection in ancient Greece was the assumption that man was unconstrained in his decision-making, thus precluding the existence of necessary laws that would rule over human intentionality. Taking various forms, this position is contained in the reflections, often contradictory, of the philosophical schools of that period, such as Epicureanism, Platonism and Aristotelianism.[1] Due to this distinction, the main areas of reflection on the social world were ethics and politics.[2] However, while Greek philosophy had a great influence on ethical ideas developed in later periods, its contribution to reliable research and understanding of the social world was scant. Consequently, as Marcin Gorazda (2014, 38) points out,

> one cannot help noticing that while in the field of metaphysics, ethics or formal sciences (mathematics and logic) the ancient thought still constitutes an important point of reference for contemporary reflection, and in natural sciences at least partly contributes to the currently used method of discovering the world, in social thought, especially in analyses of trade or the market, it remains at most a historical peculiarity.

The reflections of Greek philosophers that seem closest to the idea of spontaneous order are those from before the so-called post-Socratic period. One thinker to be named here is Hesiod, living in the 8th century BC. In his poem "Works and Days" he pointed to the spirit of rivalry born from attempts to match up with others. The rivalry, which he describes as "good conflict", made it possible to overcome the problem of the shortage of material goods needed by man.

Another thinker worth mentioning is Xenophanes (c. 570 BC–c. 475 BC). While known mainly for his application of a critical approach of rationality to the question of the divine, and in particular for his criticism of the tendency towards the anthropomorphic perception of gods, he owes his position as a philosopher mainly to his epistemological reflections, in which he draws attention to the existence of human cognitive limitations. According to Xenophanes, humans have no access to knowledge about what is available to direct experience, and he emphasises that this knowledge is accessible only through what is observable. By highlighting the difficulty of acquiring unshakeable knowledge, he indicates that what is inaccessible to experience is only a matter of opinion. As he wrote,

> no man has seen nor will there be anyone who knows about the gods and what I say about all things; for even if, in the best case, someone happened to speak what has been brought to pass, nevertheless, he himself would not know, but opinion is ordained for all.
>
> (Curd 2011, 36)

Thus, even if it were expressed, such knowledge could not enter human consciousness, because it is illusory. One can see in these words the idea of subjective interpretation and understanding by man of phenomena which, creating this illusion, make it impossible to get to the truth.

The approach presented by Xenophanes finds its continuation in the thought of Protagoras (c. 490 BC–c. 420 BC). One of the most famous sophists, he is known above all for his *homo mensura* principle, which literally means "man as measure". It is expressed in the famous statement according to which "A person is the measure of all things – of things that are, that they are, and of things that are not, that they are not" (Curd 2011, 146). This means that there is no absolute criterion for distinguishing between truth and falsehood, and the assessment of truthfulness depends on the individual's beliefs. Man constitutes this criterion for himself. In particular, such an approach emphasises not only the significance of the individual, but more importantly, the significance of how he or she perceives the phenomena observed. Thus, Protagoras points to the subjectivity of human experience and its role in people's understanding and evaluation of the surrounding reality.

Although we can observe some references to the idea of spontaneous order in the work of the philosophers discussed, it is difficult to find such reflections in the later, post-Socratic period of ancient Greek thought. This observation is interesting, since the post-Socratic period is regarded as distinct because of a turn in philosophy that occurred at that time. While the thought of the pre-Socratic period focused on ontology, natural sciences and mathematics, in Socrates' time there was a shift in the focus of deliberation towards ethics and social issues. It should be recalled, however, that social reality was perceived primarily in terms of ethics, without significantly contributing to its understanding.

The approach of post-Socratic philosophers to social reality is illustrated by the issue of the distinction between *physis* and *nomos*. *Nomos* means convention or law. It embraces morals, traditions and laws established by the state. And *physis* should be understood as nature, or that which has its origin in nature. This distinction was introduced in the 5th century BC, most probably by the Sophists (Meseguer 2012, 38). This was related to the perception of the institution of law as arbitrary and changeable, which led to contrasting the mutable *nomos* with permanent and unchangeable nature (Preus 2007, 181–182). This sparked off a debate on the relationship between custom and law on the one hand and nature on the other, and on the status of moral and social norms. In this approach, *physis* related to human nature without its social implications, while *nomos* meant all laws and customs affecting human nature and resulting from the existence of society (Preus, 2007, 204–205). What is important in the context of this book is that in the dichotomy created by such a juxtaposition *nomos* embraces all human activity, both intentional and unintentional, thus constituting a form of spontaneous order. The resulting lack of a distinction between planned and spontaneous order significantly influenced the perception of social reality and its identification with things resulting from conscious and planned processes.

In the post-Socratic period, the philosophical reflections of Plato (c. 427 BC–c. 347 BC) and Aristotle (384 BC–322 BC) – the two greatest philosophers of ancient Greece – are of particular importance. Along with Socrates (427 BC–347 BC), they presented an approach as remote from the idea of spontaneous order as possible. In particular, when classifying reality they adopted a dichotomous division between *physis* and *nomos*, perceiving all social phenomena as resulting from conscious creation. Such an approach was strongly in line with the epistemological fundamentalism of both philosophers,[3] manifesting itself in the tendency to build a philosophical system based on unquestionable foundations, and thus provide certain knowledge. Considering social issues mainly in the context of ethics, these thinkers believed that a socio-political order reflecting their views on ethical issues could be designed and imposed on society from above. Such an order manifested itself in the idea of a strongly centralised city-state (*polis*). Only within the political structure of the *polis* was it possible to achieve the ultimate good that is moral perfection. In this view, the citizen and the state are closely connected with each other – the moral perfection of the state is identical with moral perfection of the citizen. At the same time, this special position of the state meant that as an instrument for shaping and nurturing virtues it had a great power over individuals, controlling their upbringing and family life.

The statism of this vision is closely linked to elitism and a strongly hierarchical structure of power. Moreover, it manifests itself in a predilection for such virtues as martial arts and agriculture. This translates into an evident nostalgia of Socrates, Plato and Aristotle for the statism and totalitarianism of the Spartan political system. These philosophers also despised all forms of enrichment and profit-seeking. Gainful work and trade were, in their opinion, vulgar and contrary to the virtues, as they did not allow for a full dedication to the pursuit of moral perfection. As a result, Plato and Aristotle denied citizenship to those seeking profit and wealth, reducing them to the role of servant in relation to the state and its citizens. These views led to criticism of the political and economic system they observed in Athens, based as it was on democracy and private property. They believed that the Athenian system was a manifestation of the moral degeneration of Greek society.

While both Plato and Aristotle sought to create an ideal social order from above, the vision of the former was more radical in this respect. For Plato, the *polis* should be arranged in a strictly hierarchical way, based on class divisions. He distinguished the ruling class, divided into philosophers and guardians, and the auxiliary class of producers and soldiers. The representatives of the former were to live in a commune, without money or private property, which, according to Plato, weakened the virtues. And the producers were to serve the ruling class, providing them with material goods. Moreover, they were to be prevented from getting rich. Additionally, every form of intellectual activity was to be subject to censorship.

Aristotle was more moderate in his analyses, but he also endowed the state with a deliberate character. As he pointed out, "we do not choose everything for the sake of something else" (Aristotle 1996, 1094a), meaning that there is an ultimate goal, namely the highest good. This was only possible thanks to the *polis*, the highest and ultimate form of political structure, resulting from the natural aspiration of human nature to moral perfection. While, unlike Plato, this approach assumed the existence of some form of change and progress in relation to *nomos*, such change was teleological in its nature. Moreover, according to Aristotle in his time this process was already over, having reached its final state. As a result, the central role of the planned state allowed Aristotle to define the right age for marriage or the acceptable number of offspring. But he also defended private property and the institution of the family, arguing that as long as they served their ultimate purpose, they were natural and justified. Furthermore, like Socrates and Plato he excluded the possibility of spontaneous social phenomena. He saw the *polis* only as a top-down, self-sustaining structure (an autarky). In his opinion, there couldn't be a city-state with more than 100 thousand inhabitants, because the government would not be able to organise such a large structure. Another effect of considering all the effects of human activity as deliberate was Aristotle's belief that only the intentional and perceptible support of others was morally acceptable. In effect, he condemned the pursuit of profit as unnatural.

It should be noted that a significant feature of both Plato's and Aristotle's thought was the absence of the issue of the limits of state interference in the sphere of individual life or the problem of governmental self-restraint. This is related to the fact that neither of the philosophers perceived the state and the individual as forming an antithesis. On the contrary, both these aspects were necessarily identical in achieving the ultimate good.

Undoubtedly, the thought of Plato and Aristotle, and their miscomprehension of social phenomena, provides a peculiar contrast to the spontaneous order approach. The effects of such an attitude were concisely expressed by Murray Rothbard, who wrote: "When man turns the use of his reason from the inanimate world to man himself and to social organization, it becomes difficult for pure reason to avoid giving way to the biases and prejudices of the political framework of the age. This was all too true of the Greeks, including the Socratics, Plato and Aristotle" (Rothbard 1995, 6). At the same time, their work had an overwhelming impact on the thinking of later epochs, for example, by shaping Christian thought to a considerable extent. Thus, the approach of Greek philosophers, burdened with statism and elitism, significantly influenced the perception of social phenomena and became a long-lasting obstacle to their study (Schumpeter 1996; Rothbard 1995; Meseguer 2012). However, it should be remarked at this point that criticising Greek thinkers for their inability to study the complex reality of spontaneous processes, which, after all, developed as a theory only in the 19th century, can be seen as unjustified. Rather, the crucial task is to draw attention to their attitude, which contributed to such a situation. They were convinced that they had access to knowledge giving them the right to forcefully impose their own point of view on others.[4] This manifestation of a certain lack of intellectual modesty was described by Hayek as a fatal conceit, and he gave such a title to one of his books (Hayek 1988). This attitude is more than evident in these two philosophers' approach to economic phenomena, which they did not understand. As Jesus Huerta de Soto writes:

> From the standpoint of our topic, the principal characteristic shared by Socrates, Plato, and Aristotle – the three greatest philosophers of ancient Greece – was their inability to grasp the nature of the flourishing mercantile and commercial process taking place between the different Greek cities or *polis* (both in Greece itself and in Asia Minor and the rest of the Mediterranean). When they spoke of the economy, these philosophers relied on their instincts rather than on observation and reason. They scorned the work of craftsmen and merchants and underestimated the importance of their disciplined, daily efforts. Hence, it was through these philosophers that the traditional opposition of intellectuals to anything involving trade, industry, or entrepreneurial profit began. This "anticapitalistic mentality" would become a constant theme among "enlightened" thinkers all through the intellectual history of mankind from that point until our time.
>
> (Huerta de Soto 2010a)

This attitude, which Hayek described as "philosophers' blindness" (Hayek 1988, 45), is all the more interesting because, as residents of Athens, all three philosophers had the opportunity to observe the importance of the relative freedom of trade for the dynamic economic growth and political development of the *polis*. Around the 5th century BC, Athens had about 300 thousand inhabitants. Of these, 25,000 free foreigners and a quarter of the 160,000 free citizens were engaged in gainful employment, crafts and trade. This meant that less than half of the population of the *polis* was engaged in agricultural production, which, given the economic structure of those times, was an extraordinary achievement (Gorazda 2014, 40).

Despite the "blindness" of Greek philosophers and their ignorance of economic reality, it is remarkable that the reflections of later thinkers, and thus the development of the concept of spontaneous order, were largely based on Aristotle's philosophical inquiries, in particular on the distinction between form and matter.[5] This philosopher also distinguished exchange value from the utility value of a given good. The former meant the market price expressed in monetary units, while the latter contained the estimation of the usefulness of a good by an individual, thus giving it a subjective character. While this did not translate into Aristotle's understanding of market mechanisms, the concept itself was adopted and developed by later Christian thinkers, including St. Thomas Aquinas, which in the late Spanish scholastics led to the emergence of the subjective value theory.[6]

The approach to social reality slightly changed in the 3rd century BC thanks to the Stoics. They developed and systematised the concept of natural law, conceived as absolute and universal, encompassing all people and thus standing above legislated law. This idea, invoking the concept of *physis*, differed from previous considerations on this subject. Unlike the Sophists, the Stoics did not see *physis* as a state reducing the relationship between people to the principle of the power of the stronger over the weaker. At the same time, unlike the philosophers discussed earlier, whose approach was closely related to the idea of the *polis*, they focused on the individual, not the state, in their reflections on this law. For natural law can be discovered by human reason. As a result, as Rothbard (1995, 21) noted, this approach had important political implications, for the first time forming the basis for moral and transcendent criticism of positive law from the standpoint of universal human nature. The ideas of the Stoic school were gradually transferred from Greece to Rome, strongly affecting Roman jurists in the 2nd and 3rd centuries AD.

It should be noted that, regardless of the influence of the Stoics and the concept of natural law on Roman law, there are clear references to the idea of spontaneous order in the legal and philosophical deliberations of the ancient Romans. They stem from the nature of Roman civil law (*ius civile*). This was not a form of statute law, but was created by jurists seeking solutions to specific problems submitted to them by citizens. While the Romans paid little attention to reflecting on social or economic reality, they were aware of the unplanned nature of their law. Representatives of classical Roman jurisprudence pointed out that the law was not the result of the intentional actions of specific people, but a joint work of subsequent generations, developed over the course of centuries, representing knowledge going beyond the cognitive capabilities of man. Such an approach perfectly reflects the claim attributed by Cicero (106 BC–43 BC) to Cato the Elder (234 BC–149 BC):

> Cato used to say that the organisation of our state surpassed all other states for this reason: in others there were generally single individuals who had set up the laws and institutions of their commonwealths – Minos in Crete, Lycurgus in Sparta [...]. Our commonwealth, in contrast, was shaped not by one man's talent but by that of many; and not in one person's lifetime, but over many generations. He said there never was a genius so great that he could miss nothing, nor could all the geniuses in the world brought together in one place at one time foresee all contingencies without the practical experience afforded by the passage of time.
>
> (Cicero 2017, 33)

The aim of the jurists was to search for universal legal rules and to apply them to solve particular problems (Huerta de Soto 2009, 24–25). Thus, as Bruno Leoni pointed out (1961, 84), the creation of law was not the result of its deliberate establishment, but of discovering and describing it. This meant that it was seen as a common heritage that no one could change at will.

The scholastics

The notion of subjectivity, crucial for the idea of spontaneous order, can be found in medieval and late scholastic works. It appeared mainly in reflections on just price (*iustum pretium*). Since the 13th century, with the development of Thomism, they were dominated by an approach according to which the price did not lie in the essence of a given good, but resulted from its usefulness and human needs (Rothbard, 1995, 31–94). The discovery and partial reception of Aristotle's thoughts by St. Albert the Great (c. 1200–1280), and in particular by his disciple, St. Thomas Aquinas (1225–1274), had a significant impact on this approach. The latter adopted an Aristotelian perspective, according to which the exchange value of a good was determined by its usefulness and the need of the consumer. Apart from Aquinas, such an approach is visible in Giles of Lessines (d. 1305) and Richard of Middleton (c. 1249–1306). And Peter John Olivi (1248–1298) defined subjective utility (*complacibilitas*). While he pointed out that the price of goods and services resulted from a common estimation made by members of the community, by this he meant market valuation, not top-down regulation. As he wrote, the just price is "the one which happens to prevail at a given time according to the estimation of the market, that is, what the commodities for sale are then commonly worth in a certain place" (Roover 1967, 20). This concept was later also discussed by St. Bernard of Siena (1380–1444) and St. Anthony of Florence (1389–1459). Similar reflections can be found in the writings of John Buridan (1300–1358), who showed that widespread need and usefulness led to the emergence of a market price. Moreover, this philosopher was probably the first to note that subjective preferences were revealed in the process of voluntary exchange (Rothbard 2006). At the same time, he rejected the belief, going back to Aristotle and still maintained by Aquinas and others, that money was a form of convention deliberately established by people. Buridan treated money as a market phenomenon. He claimed that it was created in a natural way, as a market commodity. It was the market itself that chose the means of exchange, preferring metals as having the best qualities to play the role of money (Rothbard 1995, 73).

The observation that a just, and therefore natural, price did not result from the nature of the thing itself, but from a common valuation, led the scholastics to recognise a category of phenomena which were the work of man, but did not result from their deliberate creation. This was an intermediate category between natural phenomena (*physis*) and what resulted from intention or convention (*nomos*). Beginning in the 12th century, some medieval thinkers began to include such intermediate categories in the category of natural phenomena (*naturalis*), thus departing from the order harking back to ancient Greece (Hayek 1973, 21). At the same time, in the Middle Ages, there was a change in the perception of gainful employment and trade. For the ancient Greeks, most of whom never had to take up work to ensure their livelihood, all forms of work were unworthy of the citizen and contrary to the virtues. This approach contrasted with the much greater egalitarianism of early Christianity and its attractiveness to the middle class getting their living from work and trade. This egalitarianism was naturally reflected in the origins of Christian thinkers, who often came from less privileged social strata, which in turn influenced their perception of work.[7] Along with the praise for work contained in the Old Testament, taken over by Christianity, this contributed to endowing work with a positive aura as one of the factors in the development of society. As for trade, the process of change in the prevailing attitudes took more time. It was believed that buying cheaply and selling dearly – which is the essence of trade – constituted "shameful gain" (*turpe lucrum*) and involved fraud, and therefore was sinful. A re-assessment of such an approach to trade only took place with the economic recovery and the development of scholastic teaching at the beginning of the second millennium. Consequently, the importance of trade began to be appreciated. As Thomas of Chobham (1160– c. 1233) wrote, by supplying goods, merchants contributed to the better satisfaction of human needs. St. Albert the Great, Peter of Tarentaise (c. 1224–1276, later Pope Innocent V) and St. Bonaventure (1221–1274), also stressed the fundamental importance of merchants for society.

The subjective approach and change in the perception of the category of natural phenomena was to the largest extent developed by 16th-century Spanish scholastics. This resulted from a rather surprising, given the disappearance of scholasticism at the end of the Middle Ages, revival of this school in Spain in the 16th and 17th centuries, during the so-called "Spanish Golden Age" (*Siglo de Oro Español*). Since the majority of Spanish thinkers of that period, mainly Jesuits, were associated with the University of Salamanca, this group became collectively known as the Salamanca School (Huerta de Soto 2010b, 204). While in the matter of natural law this school reproduced the earlier conclusions of the Thomists, it greatly developed reflection on the market. Spanish scholastics, while maintaining the division of phenomena into natural and artificial, classified all phenomena not resulting from intended creation as natural. In particular, Luis de Molina (1535–1600) indicated that the natural price was natural regardless of law or decree. It resulted from a number of circumstances derived from the estimation made by individuals (Hayek 1973, 21). The subjective nature of this estimation was pointed out, among others, by Diego de Covarrubias y Leyva (1512–1577), who expressed the essence of this matter in the following way: "The value of an article does not depend on its essential nature, but on the estimation of men, even if that estimation be foolish. Thus, in the Indies wheat is dearer than in Spain because men esteem it more highly, though the nature of the wheat is the same in both places" (Grice-Hutchinson 1952, 48). The same approach can be found in Juan de Mariana (1536–1623) and Luis Saravia de la Calle (16th century), who was the first to point out that it was the price that determined costs and not the other way around (Huerta de Soto 2008, 30). At the same time, understanding the factors behind price formation allowed Martín de Azpilcueta (1491–1586) to observe that this mechanism should not be manipulated by the State, and pointed to the negative effect of such actions. As Rothbard wrote, "Azpilcueta was the first economic thinker to state clearly and boldly that government price-fixing was imprudent and unwise. When goods are abundant, he sensibly pointed out, there is no need for maximum price control, and when goods are scarce, controls would do the community more harm than good" (Rothbard 1995, 105).

The subjective value theory described by Spanish scholastics allowed for an understanding of the complexity of the price setting mechanism itself. As

Juan de Salas (1553–1612) and Juan de Lugo (1583–1660) noted, the number of circumstances affecting the price is so large that it makes acquiring complete knowledge of these circumstances impossible. The problem of insufficient information was also raised by Juan de Mariana, but he went beyond the market mechanisms in his analysis. In his work *A Discourse on the Sicknesses of the Jesuit Order*,[8] he criticised the military character of the Order's structures. Importantly, de Mariana also spoke about the actions of the state, pointing to the negative effects of the centralisation of power. The Jesuit believed that the government was unable to acquire all the information needed to organise society. As he wrote about the government, "[…] it knows neither people nor events, at least in relation to the circumstances on which success depends. […] How can the government function properly without knowing about everything and everyone? Inevitably, it will make a lot of serious mistakes that people will be concerned about and disregard such a blind government."[9] But when the government is unable to achieve this goal, its actions lead to a violation of the existing order. For by trying to manage the community from above, the government creates a situation in which "there are too many laws […] and as not all of them can be respected, or even known, respect for all of them disappears".[10]

The Enlightenment

In the 17th century, the importance of the scholastic school started to decline, which was associated, among other things, with the development of Cartesian rationalism and Baconian empiricism. Moreover, the Thomism of the Scholastics, together with the entire philosophy of the Roman Catholic Church, was criticised by Protestants. As a result, the thought of the late scholastics remained unnoticed in Britain. However, the tradition of the scholastic school did not disappear altogether, staying relatively strong in the Catholic countries. The subjectivism of the Salamanca School was adopted, for example, by Italian and French thinkers. But the question of human limitations in creating social order was discussed also in Britain. This was related to the development of reflection on precedent-based law. Like Roman civil law, this was not a form of statutory law but was created by drawing on tradition and custom in relation to the specific circumstances of particular cases. This system was strongly influenced by Matthew Hale (1609–1676). Pointing to the complexity of the law and at human ignorance, he criticised the approach saying that law could be made on the basis of reason. He believed it was "a foolish and unreasonable thing for any to find fault with an institution because he thinks he could have made a better, or expect a mathematical demonstration to evince the reasonableness of an institution or the selfe evidence thereof" (Hale 1924, 505). Since the law is the work of many people over many centuries, it is not possible for a human being to fully comprehend this institution. Attempts to change it arbitrarily would undermine its certainty and lead to unintended harm. Only within the framework of precedent-based law, drawing on the fount of specific cases, is it possible to preserve its uniformity, predictability and certainty, and to avoid making arbitrary decisions. Consequently, as Hale wrote, "It is a reason for me to preferre a law by which a kingdome hath been happily governed four or five hundred years than to adventure the happiness and peace of a kingdome upon some new theory of my own" (Hale 1924, 504).

In the 18th century, unplanned order was analysed by representatives of the Scottish Enlightenment. The precursor of this trend was Bernard Mandeville (1670–1733). In his best known work, *The Fable of the Bees: or, Private Vices, Public Benefits*,[11] he describes a community of bees living affluently, but in contravention to moral principles, which by way of Jupiter's intervention are transformed into moral beings and purged of all vices. However, the absence of crime or greed leads to the collapse of the economy and to unemployment of those who previously owed their jobs to the immoral behaviour of others – judges or providers of goods and services to the rich. Mandeville points out that paradoxically it is human weaknesses and selfish conduct that are the main cause of shared prosperity. Virtue alone cannot deliver wealth and only leads to poverty. He also points out that people's selfish conduct must be institutionally constrained. Their orientation towards the common good is not a self-regulating system, but results from political decisions.[12]

Recognising the existence of unplanned structures, Mandeville seems to be aware of the limitations of human knowledge. Arguing with Hale, he opposed the perception of laws as having their origin in independent reason. As he pointed out, "That we often ascribe to the Excellency of Man's Genius, and the Depth of his Penetration, what is in Reality owing to length of Time, and the Experience of many Generations, all of them very little differing from one another in natural Parts and Sagacity" (Mandeville 1924, 142). In a similar vein, he also pointed to the evolutionary nature of language, at a time when it was still perceived as given to man (Fitch 2010, 438).

Challenging as it did the belief in the existence of the divine order, Mandeville's work provoked a wide-ranging, mostly critical response, which made it extremely popular (Mandeville 1924, cxiv–cxlvi).[13] In particular, references to it can be found in the works of other representatives of the Scottish Enlightenment, who also dealt with the issue of actions of individual people pursuing their own goals as leading to the common good. However, they rejected the claim that the common good may result from immoral and illegal conduct. David Hume (1711–1776), sceptical of human knowledge, pointed out the limitations of the mind in explaining and justifying moral judgments, which meant that moral and legal norms serving society could not be defined by reason alone. Discovering them was only possible by drawing on tradition and experience. Hume assumed a uniformity of human nature, implying the existence of a permanent structure of general social principles, reflected in prevailing norms. This anti-rationalist approach found its expression in a criticism of the contractualist approach, seeing society as a product of a rationally formulated contract. The critics of this view instead proposed that sources of law and order were to be found in man's natural inclinations, including the pursuit of personal gain. In the context of the common good, Hume wrote:

> 'Tis self-love which is their real origin; and as the self-love of one person is naturally contrary to that of another, these several interested passions are oblig'd to adjust themselves after such a manner as to concur in some system of conduct and behaviour. This system, therefore, comprehending the interest of each individual, is of course advantageous to the public; tho' it be not intended for that purpose by the inventors.
>
> (Hume 2007, 339)

Thus, he indicated the existence of a spontaneous process of interaction between people oriented towards their own goals leading to the emergence of social order.

In a similar vein, Adam Ferguson (1723–1816) also stressed the limitations of reason, as significantly affecting the perception of human nature as closely related to morality. Because society, and the moral norms that exist in it, is not an effect of a rational plan, but of human nature. And man, although bent mainly on his own survival and interest, is also capable of creating positive norms and charity. The pursuit of one's own goals does not preclude promoting the common good. The unplanned nature of social reality, resulting from the actions of individuals, was concisely expressed by Ferguson in his work *Essay on the History of Civil Society*,[14] where he said that "Every step and every movement of the multitude, even in what are termed enlightened ages, are made with equal blindness to the future; and nations stumble upon establishments, which are indeed the result of human action, but not the execution of any human design" (Ferguson 1966, 122).

Adam Smith (1723–1790) is one of the most recognised representatives of the Scottish Enlightenment. He also analysed the question of a process in which self-interested individuals benefit others through their actions. While this had already been explained by Mandeville and Ferguson, Smith's main merit seems to be the creation of the explanatory metaphor of an invisible hand leading man to promote unintended goals. He wrote:

> As every individual, therefore, endeavours as much as he can both to employ his capital in the support of domestic industry, and so to direct that industry that its produce may be of the greatest value; every individual necessarily labours to render the annual revenue of the society as great as he can. He generally, indeed, neither intends to promote the public interest, nor knows how much he is promoting it. By preferring the support of domestic to that of foreign industry, he intends only his own security; and by directing that industry in such a manner as its produce may be of the greatest value, he intends only his own gain, and he is in this, as in many other cases, led by an invisible hand to promote an end which was no part of his intention. Nor is it always the worse for the society that it was no part of it. By pursuing his own interest he frequently promotes that of the society more effectually than when he really intends to promote it.
>
> (Smith 1979, 456)

However, Smith pointed out that the invisible hand is not independent and must be placed within a fair system of law that enables individuals' own interests to be channelled towards the common good and which is a prerequisite of the market. He also indicated that apart from self-interest, the motivation to act is also influenced by moral feelings, allowing people to pass moral judgements. These feelings decline in importance once interactions between people become increasingly anonymous and less personal. Smith also attached great importance to the issue of the division of labour, which is an expression of cooperation between people resulting from the operation of the invisible hand.

Smith is a controversial figure. On the one hand, his social reflection is the most systematic among the Scottish Enlightenment thinkers – a group identified with the origins of the tradition of spontaneous order (Barry 1982, 25; Horwitz 2001, 81–83). On the other hand, he is criticised by some contemporary researchers of the history of economic thought, who accuse him of amply and often inaccurately borrowing from the thought of his contemporaries and plagiarising their works (Rothbard 1995; Meseguer 2012; Huerta de Soto 2008). Due to the popularity of Smith's reflections, this even resulted in his work being considered retrograde to the development of economic thought. As Rothbard wrote:

> [T]he most malignant aspect of this Smith-worship is that the lost economists were in many respects far sounder than Adam Smith, and in forgetting them, much of sound economics was lost for at least a century. In many ways [...] Adam Smith deflected economics, the economics of the Continental tradition beginning with the medieval and later scholastics and continuing through French and Italian writers of the eighteenth century, from a correct path, and on to a very different and fallacious one. Smithian "classical economics", as we have come to call it, was mired in aggregative analysis, cost-of-production theory of value, static equilibrium states, artificial division into "micro" and "macro", and an entire baggage of holistic and static analysis.
>
> (Rothbard 1995, 361)

From the point of view of this book, it is important to note that while Smith and other Scottish Enlightenment thinkers recognised the existence of spontaneous social processes, their understanding of these processes was based on classical economics. The question of value, according to Smith determined by the cost of labour, is particularly relevant here. Such an approach posed an obstacle to understanding spontaneous processes of social order formation, because Scottish Enlightenment thinkers ignored the subjective nature of value.

However, while the theory of subjective value developed by the scholastics did not make a significant impact in Britain, it played an important role in continental thought.[15] Richard Cantillon (c. 1680–1734) deserves attention among the thinkers adopting this approach, clearly seen in the initial words of his only surviving work, "Essay on the Nature of Trade in General",[16] where he defined wealth as "nothing more than food, amenities and pleasures of life".[17] He indicated that the value of goods depended on supply and demand, influenced in its turn by "people's moods and fantasies, and consumption".[18] The subjective nature of people's estimation of value is linked to Cantillon's perception that market activities take place under conditions of permanent uncertainty. This leads to a crucial role for entrepreneurs, whose work consists in taking risks. This approach differs significantly from the conception of Smith, who did not see the importance of entrepreneurs in economic processes. The differences between the two thinkers are also visible in their perception of spontaneous market processes. Both Cantillon and Smith were aware of the negative effects of state interference in the economy. They also saw the existence of a certain mechanism of self-regulation, ensuring social harmony. The Scottish philosopher explained this through the idea of the invisible hand. However, the invisible hand does not have its origin in man, but in the natural order of things, and getting to know it is beyond human abilities. Cantillon, on the other hand, is closer to the idea of methodological individualism, claiming that the creation of order is based on decisions made by individual people, influenced by many factors including social and cultural circumstances.

Cantillon's approach to studying economics led him to focus on the complexity of economic processes. He was the first economist to use a thought experiment, the idea of which is to abstract the studied phenomenon from a complex reality and then idealise it (Rothbard 1995, 348). The model created in this way, through reasoning based on the principles of logic (deductive reasoning), serves to draw conclusions about real phenomena. As a result, Cantillon became the first to apply the a priori and deductive method (Gorazda 2014, 120). Furthermore, he was the first to outline the boundaries of a new science, which, together with the systematic nature of his thought, led William Jevons (1881, 342) to describe his work as "the first treatise on economics" (Jevons 1881, 342). Despite the fact that Smith is widely recognised as the father of modern economics, some historians of economics, especially those associated with the Austrian School, propose that this title should go to Cantillon (Gorazda 2014, 106).

Cantillon had a strong influence on the French physiocrats, their main representative being François Quesnay (1694–1774). They emphasised the self-regulatory nature of market mechanisms, meaning that they should not be subject to intervention. Asked what the government's economic policy should be, Vincent de Gournay (1712–1759) supposedly answered with the words *laissez faire, laissez passer* (let do, let pass), giving rise to the concept of laissez-faire. Anne-Robert-Jacques Turgot (1727–1781), in praising commercial freedom, also pointed to the benefit of man's pursuit of his own well-being for the common good. Moreover, we can find a subjective approach to the concept of value in his reflections. He also noted that each person has specific knowledge, acquired through his actions in the process of trial and error. The resulting dispersion of knowledge precludes the top-down planning and management of market processes. As Turgot pointed out, economic freedom is based on "the absolute impossibility of managing, through constant rules and constant control, many transactions whose sheer multitude makes it impossible to know them fully, and which depend on many constantly changing circumstances, which cannot be controlled or predicted".[19]

The Austrian School of Economics

A key figure in the development of the theory of spontaneous order was Carl Menger (1840–1921). He was the first to formulate a comprehensive theory describing the process of the spontaneous formation of social institutions (Huerta de Soto 2008, 37). Menger based his analyses on the premise of methodological individualism, considering man to be the fundamental causative factor of social phenomena. However, the key element of his thought was a subjective notion of human conduct, deriving from subjective value theory. This is based on the subjective perception of reality by man, which means that man acts by pursuing a goal to which he has assigned a certain subjective value. This allowed Menger to create a "composite" – to use his own term – method of explaining social phenomena. According to the Austrian, such an explanation, which Mayer (1994, 57) defines as genetic-causal, is only possible through reference to the actions of individuals. It requires reconstructing the process that led to the creation of the phenomenon being explained.[20] Thus, he rejected a functional approach based on the intentional perceptions of institutions.

Methodological subjectivism, along with the use of the composite method, emphasised the importance of the unintended consequences of human actions. As a result, Menger combined his approach with the introduction of a distinction between "organic" institutions, which were "the unintended result of human efforts aimed at achieving essentially individual goals" and "pragmatic" ones, established deliberately (Menger 1985, 133). He stressed that we cannot attribute any design to many institutions contributing to the common good: "Language, religion, law, even the state itself, and to mention a few economic social phenomena, the phenomena of markets, of competition, of money, and numerous other social structures are already met with in epochs of history where we cannot properly speak of purposeful activity of the community as such directed at establishing them" (Menger 1985, 146). At the same time, in keeping with the traditions of the 18th-century representatives of classical liberalism, Menger wondered how organic institutions could serve universal welfare.

A classic example of Menger's use of the genetic-causal method is his explanation of the emergence of the institution of money. As he pointed out, the belief in the deliberate creation of money stemmed from an attempt to avoid a certain paradox associated with it. If one assumes that people acting in their own interest exchange only the goods they need, it seems impossible to create money or any other institution that they do not directly need without some form of informed consent or a top-down order.

Menger pointed out that the institution of money results from an unplanned process. Considering a situation in which due to the absence of the institution of money only barter exchange functions, he noted that in such circumstances exchange of goods for consumption can only take place if both sides want each other's goods. In the absence of such a coincidence, a person is not able to exchange his or her goods for ones he or she wants. Consequently, some people begin to seek an intermediate good for which they can buy the desired goods. This leads to the emergence of goods that are exchangeable for more goods than others. People begin to exchange their less exchangeable goods for more exchangeable goods, even though they do not necessarily need them directly. Ultimately, this process results in the emergence of a small number of the most liquid (preferably exchangeable) goods, which become a commonly accepted means of exchange, that is, money.

Menger is regarded as the founder of the Austrian School of Economics. The adjective "Austrian" refers to the views of his and his successors, also connected with Vienna. It was given by representatives of the German Historical School in the context of a dispute between the two groups (Mises 1984, 19).[21] This debate, called *Methodenstreit*, regarded the question of methods of studying economic processes.[22] It seems, however, that from the economic standpoint, Menger's more widely recognised achievement is his formulation of the marginalist theory of utility. It says that the value of a good is determined by the subjective estimation of the value of the last consumed unit of a given good. Together with analogous theories formulated by William Jevons (1835–1882) and Léon Walras (1834–1910) in the same period, this made it possible to overcome the domination, initiated by Smith, of the British school of classical economics, based on an objective understanding of value. This change is called the marginalist revolution.

It should be stressed that Menger's approach to the issue of value, and to human action in general, was part of a subjective tradition that had survived on the continent despite the popularity of the Anglo-Saxon approach. This tradition, associated with Aristotle's philosophy, functioned mainly in France, Italy and Spain (Rothbard 2006).[23] In particular, this led Menger to rediscover the thought of the Spanish scholastics (Huerta de Soto 2010b, 210). In this context, the impact of Aristotle's philosophy, but also of Thomism, on the intellectual climate in Austria was undoubtedly an important factor contributing to the development of the subjective concept of human action by Menger and his followers (Kauder 1958, 419–420). Subjectivism, and in particular Menger's drawing on the traditions of the Spanish scholastics, as suggested by Huerta de Soto (2008, 34), was also associated with the strong political and cultural relations that had developed between Spain and Austria since the 16th century. This was reflected in the difference between the Austrian School and other marginalist approaches. Unlike Jevos and Warlas, who used the Aristotelian distinction between form and matter, the Austrians believed that economics explored not quantitative phenomena but the essence of real phenomena such as value or profit. This shaped the perception of these phenomena in terms of Menger's subjective concept of human action. The theory of marginal utility itself was only a conclusion naturally following from the application of this concept.[24]

The approach proposed by Menger was developed by successive representatives of the Austrian School of Economics. Ludwig von Mises (1881–1973) and Friedrich Hayek (1899–1992) particularly contributed to the development of the theory of spontaneous order. The first of these thinkers is considered the most important economist of the Austrian School in the 20th century (Hülsmann 2007). Mises took the concept of human action as the starting point of his reflections. This meant that although he focused mainly on the study of economic processes, he considered economics to be only one of the branches of the more general science of human action, known as praxeology.[25] This approach was closely related to methodological subjectivism, since in the view of the Austrians human conduct was based on a subjective evaluation of objectives and the means to achieve them. As a result, Mises created the theory of the impossibility of socialism. In this theory, he pointed out that market pricing results from an interpersonal exchange of goods based on a subjective estimation of their value by individual actors. The coercion ingrained in socialism, and the interventionist system, constrain free exchange, preventing the emergence of a price system. Without this information, the central decision-making body is not able to make an economic calculation, conceived as an assessment of the value of alternative decisions. Consequently, it is not able to determine whether the measures taken are profitable or not.

Hayek's achievements are mainly connected with economics, for which he was awarded the Nobel Prize in 1974. However, he also made important contributions in areas such as linguistics, cognitive science and philosophy of science. In the context of the theory of spontaneous order, of particular importance are his reflections on the epistemic limitations of human ability to plan the creation and management of social institutions. The source of these limitations is the dispersion of knowledge, of which each person has a unique portion. This knowledge becomes accessible to other people through the social institutions possessing it. This approach was a generalisation of Mises' reflections on the institution of price. Like Mises, Hayek pointed out that no central supervisory body is capable of collecting all information. He also strongly criticised attempts to control society from above, in which he saw a manifestation of the fatal conceit of reason. In addition, Hayek argued that not only information is dispersed, but also that not all of it can be expressed in some kind of language. Part of human knowledge is of an inarticulate nature, which poses another obstacle to planned institution-building.

It is worth noting that Hayek's focus on knowledge and the availability of information led him to move away from perceiving the causes of the creation of spontaneous order in the manner of 17th- and 18th-century thinkers like Smith. The Austrian pointed out that lack of coordination results not so much from selfishness as from the inevitable ignorance of the individuals involved.

While the concept of spontaneous order is commonly associated with Hayek, the first use of the term "spontaneous order" is attributed to Wilhelm Röpke (1899–1966), a representative of ordoliberalism (Bladel 2005, 23). However, the key figures from outside the circle of the Austrian School of Economics dealing with the issue of spontaneous order were Bruno Leoni (1913–1967) and Michael Polanyi (1891–1976). Lon Fuller's (1902–1978) reflections on the philosophy of law also deserve mention.

Leoni, not being an economist but an expert on law, focused on the issue of the emergence of the legal order. In particular, he analysed limitations in the planned formation of this order, and thus also the social order, indicating that systems with a central decision-making body lack the knowledge necessary for managing them. He perceived the law as a set of common rules arising in the dynamic process of discovering solutions to problems by individual participants of this system. He pointed out that law should not be constructed, but discovered within a polycentric system, emphasising the importance of the decentralised system of Roman law and English precedent law for the concept of the rule of law.[26] In such a system, decisions can be made only by the parties concerned and are usually linked to existing case-law. This provides for the certainty of the law formed in this way. Leoni saw a strong affinity between such a polycentric system of law and cooperation in the free market. He also criticised the institution of statutory law as an example of central control of a structure characterised by a complexity going beyond human cognitive abilities. The spread of such mechanisms is associated with the belief in the human ability of deliberate shaping the social order and making the law teleological. This means that it ceases to be a process of discovery. These observations led the Italian researcher to conclude that his approach was a generalisation of Mises's criticism of central planning in the economy. As he wrote:

> The fact that the central authorities in a totalitarian economy lack any knowledge of market prices in making their economic plans is only a corollary of the fact that central authorities always lack a sufficient knowledge of the infinite number of elements and factors that contribute to the social intercourse of individuals at any time and at any level. The authorities can never be certain that what they do is actually what people would like them to do, just as people can never be certain that what they want to do will not be interfered with by the authorities if the latter are to direct the whole law-making process of the country.
>
> (Leoni 1961, 90)

Fuller presented views close to Leoni's approach. This representative of legal naturalism perceived law as an immanent part of nature. Therefore, making law was for him not a process of its arbitrary creation, but of discovering the rules of its formation.[27] Like Leoni, he noticed an affinity between law and the free market, based on reciprocity and equality. In particular, both institutions have a coordinating function, enabling the participants to plan their future activities. He also stressed the danger inherent in statutory law and the legislative process. Justifying this position, he pointed to human cognitive limitations and the resulting flaw in the planned creation of the legal system. "No single concentration of intelligence, insight, and good will, however strategically located, can insure the success of the enterprise of subjecting human conduct to the governance of rules" (Fuller 1964, 91). This means that law should not be seen in terms of the deliberate creation of an ideal social order, but as an institutional framework enabling individuals to realise their creative potential. This approach is reflected in Fuller's distinction between the morality of duty and the morality of aspiration. The former defines rules necessary for the functioning of society, the breaking of which is unacceptable. The latter reflects the ethical ideas man should pursue. Fuller warns against including the morality of aspiration in the law, because due to human cognitive limitations, it leads to limiting human creativity and may undermine the coordinating role of the law.

Just like the more famous Hayek, Polanyi also worked on the idea of self-organising structures. He introduced the distinction between corporate and dynamic order. The former is exogenous, meaning that relations between its elements are determined by external factors. In the endogenous dynamic orders, on the other hand, the behaviour of a given element depends on the behaviour of other elements. Thus, regularities defining a system result from a process in which individual elements adapt their behaviour to each other. Such a system has no superior body imposing its decisions on it. Thanks to such polycentrism individuals living in social orders are free to act in the way they consider best, using their personal and unique knowledge. As Polanyi pointed out, such spontaneity is particularly appropriate for the free market economy, but also for the law or the world of science. Speaking about all polycentric systems, he stressed the key role of the process of discovery and individual knowledge of particular actors. Additionally, self-organised systems of that type are often characterised by a high degree of complexity.

Notes

1 A reference to necessary laws can in turn be found in the theme of *fatum*. This word means blind fate or destiny, but it also invokes necessary law. However, the reference to the notion of a necessary law or destiny contained in this world is a form of reference to the supernatural. Thus, it is remote from the idea of a universal and cognitive law governing the social world.

2 It should be noted that the ancient Greeks saw the laws of physics and moral rules as ontologically related. Both belonged to the rational structure of the world (*logos*). This meant that while man was not subject to the necessary laws in his actions, he was expected to act in accordance with the *logos* and the moral rules contained in it. The unity of the structure of the world (of its physical and moral aspect) perceived in this way meant the lack of a clear separation between what "is" and what "should" be (Brożek 2016, 217–223).

3 This approach is also sometimes described as foundationalism (see Grobler 2006, 68–69).

4 In Plato this approach does not remain in the verbal sphere, but also leads to political action (Huerta de Soto 2010a). In Syracuse he tried, together with his students, to put the vision of a utopian tyrant into practice. And in Athens Antipater and Demetrius of Phalerum (Barker 1958, xxiv–xxvi) used Aristotle's vision of the *polis* in their policies.

5 The impact of Aristotle's epistemological concepts is discussed by the Austrian School of Economics.
6 More on scholastic considerations can be found in the next subsection. Aristotle's reflections on epistemology, and especially the distinction between form and matter, are also reflected in the approach of the economists of the Austrian School addressed under Section "The Austrian School of Economics" of this chapter.
7 In particular, although Christian society gradually established class divisions too, universities and monasteries were accessible to all sections of society regardless of financial issues.
8 "Discurso sobre las enfermedas de la Compañia" (Mariana 1768).
9 "no conoce las personas, ni los hechos, a lo menos, con todas las circunstancias que tienen, de que pende el acierto....? 'Pues como puede ir bien enderezado el Gobierno particular sin noticia de todo, y de todos? Forzoso es se caiga en yerros muchos, y graves, y por elos se disguste la gente, y menosprécie gobierno tanciego" (Mariana 1768, 152–153).
10 "las leyes [...] son muchas en demasía; y como no todas se pueden guardar, ni aun saber, a todas se pierde el respéto" (Mariana 1768, 216).
11 *The Fable of The Bees: or, Private Vices, Public Benefits* (Mandeville 1924).
12 The question of the extent to which Mandeville saw the benefits of human vices as requiring planned management through the political system is presented by such authors as Thomas Horne (1978) and Maurice Goldsmith (1976).
13 The ideas contained in Mandeville's work were discussed by George Berkeley, David Hume or Jean-Jacques Rousseau. The fact that between 1714 and 1732 the work run through seven editions (Mandeville 1924, ix) testifies to its popularity.
14 "An Essay on the History of Civil Society" (Ferguson 1966).
15 Objectivism in the understanding of values became one of the factors distinguishing Anglo-Saxon classical economics from approaches developed on the Continent.
16 Essai sur la nature du commerce en general (Cantillon 1959).
17 "la Richesse en elle-même, n'est autre chose que la nourriture, les commodites et les agremens de la vie" (Cantillon 1959, 2).
18 "des humeurs et des fantaisies des hommes, et de la consommation" (Cantillon 1959, 28).
19 "l'impossibilité absolue de diriger par des règles constantes et par une inspection continuelle une multitude d'opérations que leur immensité seule empêcherait de connaître, et qui de plus dépendent continuellement d'une foule de circonstances toujours changeantes, qu'on ne peut ni maîtriser ni même prévoir" (Turgot 1844, 288).
20 For the description of the composite method, see Section "Between atomism and holism" of Chapter 3.
21 In the late 1930s, with the emigration of the representatives of this approach, Vienna, and consequently Austria, ceased to be the main centre of the Austrian School of Economics. So the word "Austrian" in the name is today only a historical reference to the birthplace of this school.
22 While there is no consensus on the importance of the dispute for the development of economics, it did have an impact on the development and clarification of the position of the Austrians (Gorazda 2014, 20–22, 218; Huerta de Soto 2008, 43). The debate with the German Historical School was the first in a series of methodological disputes between representatives of the Austrian School of Economics and various schools and approaches present in the economic sciences. Consequently,

the term *Methodenstreit* embraces all these disputes (Block 2007; Huerta de Soto 2010b, 43).

23 As Emil Kauder (1958) suggested, an important factor that contributed to the distinction between the continental tradition of subjective perception of values and Anglo-Saxon objectivity was the issue of religion. He speculated that the Calvinistic approach, sanctifying and glorifying work as such, contributed to the development of the labour theory of value that emerged from the work of British thinkers. This corresponds to Kauder's observation that before the 18th century the subjective value theory was promoted only by Catholics in France and Italy. This went back to the strong presence of the thought of Aristotle and the Thomists in the Catholic tradition of continental countries, with its premise that the basis for human actions is not work, but a modest search for happiness. So it emphasises the role of the consumer and his subjective estimation of values.

24 This not only leads to a different perception of marginal utility, but also undermines the claim that the formulation of the theory of marginal utility was the fundamental achievement of Menger (Jaffé 1976; Machaj 2015).

25 Alfred Espinas is commonly regarded as the author of the concept of praxeology conceived as the science of human action (Alexandre 2000, 7).

26 The concept of the rule of law is discussed in Section "The theory of spontaneous order in relation to the epistemic system of contemporary political science" of Chapter 5.

27 Contrary to the traditional approach of the law of nature, the concept presented by Fuller did not assume the embedding of the law of nature in a transcendental being (God), but saw its origins in man himself. Thus natural law is not immutable and timeless, but subject to change in the course of social processes (Fuller 1964, 96).

Bibliography

Alexandre, Victor. 2000. Introduction. In *The Roots of Praxiology: French Action Theory from Bourdeau and Espinas to Present Days*, edited by Victor Alexandre, and Wojciech W. Gasparski, 7–20. New Brunswick: Transaction Publishers.

Aristotle. 1996. *The Works of Aristotle*. Vol. 2. Chicago: Encyclopaedia Britannica.

Barker, Ernest. 1958. *The Politics of Aristotle*. London: Oxford University Press.

Barry, Norman P. 1982. The Tradition of Spontaneous Order, Literature of Liberty. *A Review of Contemporary Liberal Thought* 5 (2): 7–58.

Bladel, John P. 2005. Against Polanyi-centrism: Hayek and the Re-emergence of 'Spontaneous Order'. *The Quarterly Journal of Austrian Economics* 8 (4): 15–30. doi: 10.1007/s12113-005-1001-x.

Block, Walter. 2007. Reply to Caplan on Austrian Economic Methodology." *Corporate Ownership and Control* 4 (3): 312–326. doi: 10.22495/cocv4i3c2p8.

Brożek, Bartosz. 2016. *Granice interpretacji*. Kraków: Copernicus Center Press.

Cantillon, Richard. 1959. *Essai sur la nature du commerce en general*. London: Frank Cass.

Cicero, Marcus Tullius. 2017. *On the Commonwealth and On the Laws*. 2nd ed. Translated and edited by James Zetzel. Cambridge: Cambridge University Press.

Curd, Patricia, ed. 2011. *A Presocratics Reader: Selected Fragments and Testimonia*. 2nd ed. Indianapolis: Hackett Publishing.

Ferguson, Adam. 1966. *An Essay on the History of Civil Society*. Edinburgh: Edinburgh University Press.

Fitch, W. Tecumseh. 2010. *The Evolution of Language*. Cambridge: Cambridge University Press.

Fuller, Lon L. 1964. *The Morality of Law*. New Haven, London: Yale University Press.

Goldsmith, Maurice. 1976. Public Virtues and Private Vices: and Ideologies in the Early Eighteenth Century. *Eighteenth-Century Studies* 9 (4): 477–510. doi: 10.2307/2737791.

Gorazda, Marcin. 2014. *Filozofia ekonomii*. Kraków: Copernicus Center Press.

Grice-Hutchinson, Marjorie. 1952. *The School of Salamanca: Readings in Spanish Monetary Theory, 1544–1605*. Oxford: Clarendon Press.

Grobler, Adam. 2006. *Metodologia nauk*. Kraków: Wydawnictwo Znak.

Hale, Matthew. 1924. *Reflections by Lord Chief Justice Hale on Minister Hobbes, His Dialogue of the Law*. In *A History of English Law*, edited by William Holdsworth, vol. 5, 499–513. London: Methuen.

Hayek, Friedrich A. 1973. *Rules and Order*, Vol. 1 of *Law, Legislation and Liberty*. London: Routledge.

Hayek, Friedrich A. 1988. *The Fatal Conceit: The Errors of Socialism*. London: Routledge.

Horne, Thomas A. 1978. *The Social and Political Thought of Bernard Mandeville*. London: Macmillan.

Horwitz, Steven. 2001. From Smith to Menger to Hayek Liberalism in the Spontaneous-Order Tradition. *The Independent Review* 6 (1): 81–97.

Huerta de Soto, Jesus. 2008. *The Austrian School: Market Order and Entrepreneurial Creativity*. Cheltenham, UK: Edward Elgar Publishing.

Huerta de Soto, Jesus. 2009. *Money, Bank Credit, and Economic Cycles*. 2nd ed. Auburn: Ludwig von Mises Institute.

Huerta de Soto, Jesus. 2010a. Economic Thought in Ancient Greece. *Mises Institute*. https://mises.org/library/economic-thought-ancient-greece.

Huerta de Soto, Jesus. 2010b. *The Theory of Dynamic Efficiency*. London: Routledge.

Hume, David. 2007. *A Treatise of Human Nature*. Vol. 1. Edited by David Fate Norton, and Mary J. Norton. Oxford: Clarendon Press.

Hülsmann, Jörg G. 2007. *Mises: The Last Knight of Liberalism*. Auburn: Ludwig von Mises Institute.

Jaffé, William. 1976. Menger, Jevons and Walras De-Homogenized. *Economic Inquiry* 14 (4): 511–524. doi: 10.1111/j.1465–7295.1976.tb00439.x.

Jevons, W. Stanley. 1881. Richard Cantillon and the Nationality of Political Economy. *Contemporary Review* 39: 333–360.

Kauder, Emil. 1958. Intellectual and Political Roots of the Older Austrian School. *Zeitschrift für Nationalökonomie* 7 (4): 411–425. doi: 10.1007/BF01318563.

Leoni, Bruno. 1961. *Freedom and the Law*. Princeton: D. Van Nostrand.

Machaj, Mateusz. 2015. Marginal Unit vs. Marginal Unit — Some Additional Thoughts on the Differences Between Menger, Jevons, and Walras. *Economia. Wroclaw Economic Review*, 21 (4): 9–16.

Mandeville, Bernard. 1924. *The Fable of the Bees: or, Private Vices, Publick Benefits*. Vol. 2. London: The Clarendon Press.

Mariana, Juan. de. 1768. *Discurso sobre las enfermedades de la Compaña*. Madrid: D. Gabriel Ramirez.

Mayer, Hans. 1994. The Cognitive Value of Functional Theories of Price: Critical and Positive Investigations Concerning the Price Problem. In *Classics in Austrian*

Economics: A Sampling in the History of a Tradition, edited by Israel M. Kirzner, vol. 2, 55–168. London: W. Pickering.

Menger, Carl. 1985. *Investigations into the Method of the Social Sciences with Special Reference to Economics*. New York: New York University Press.

Meseguer, César Martínez. 2012. *La teoría evolutiva de las instituciones. La perspectiva austriaca*. Madrid: Unión Editorial.

Mises, Ludwig von. 1984. *The Historical Setting of the Austrian School of Economics*. Auburn: Ludwig von Mises Institute.

Preus, Anthony. 2007. *Historical Dictionary of Ancient Greek Philosophy*. Lanham: The Scarecrow Press.

Roover, Raymond de. 1967. *San Bernardino of Siena and Sant'Antonino of Florence: The Two Great Economic Thinkers of the Middle Ages*. Boston: Kress Library of Business and Economics.

Rothbard, Murray N. 1995. *An Austrian Perspective on the History of Economic Thought: Economic Thought Before Adam Smith*, vol. 1. Auburn: Ludwig von Mises Institute.

Rothbard, Murray N. 2006. New Light on the Prehistory of the Austrian School. *Mises Institute*. https://mises.org/library/new-light-prehistory-austrianschool.

Schumpeter, Joseph A. 1996. *History of Economic Analysis*. Oxford: Oxford University Press.

Smith, Adam. 1979. *An Inquiry into the Nature and Causes of the Wealth of Nations*, Vol. 1 of *The Glasgow edition of Works and Correspondence of Adam Smith*, edited by R. H. Campbell, A. S. Skinner, and W. B. Todd. Oxford: Clarendon Press.

Turgot, Anne-Robert-Jacques. 1844. *Éloge de Gournay, Œuvres de Turgot*. Paris: Guillaumin.

Part II

The theory of spontaneous order

3 Epistemological foundations of the theory of spontaneous order

Human action

The theory of spontaneous order is based on a praxeological approach, which consists in interpreting reality in terms of human action, defined as intentional behaviour. This means that action is aimed at achieving the goals set by the acting person. In an attempt to clarify the definition of human action, Mises (2008, 11) described it in the following way: "Action is will put into operation and transformed into an agency, is aiming at ends and goals, is the ego's meaningful response to stimuli and to the conditions of its environment". This perception of action means that it is identical with a conscious human response.[1]

For human action to appear actors must feel they need to change a given state of affairs. This need is based on the fact that a person imagines another, more satisfactory state, which produces a sense of discomfort motivating this person to change the current situation. The third factor conditioning the action is the belief that undertaking it will make it possible to achieve the intended goal and thus to get rid of the discomfort. This indicates that the ultimate goal is always to satisfy some need of the acting person, and thus remove the discomfort.[2] It is worth emphasising that the process of selecting goals characterised in this way does not impose significant limitations on what the source of satisfaction or discomfort is for a given person. In particular, this approach is far from a narrow understanding of the pursuit of satisfaction as egoistic conduct aimed only at maximising one's own interest and to improve one's own situation. Altruistic behaviour can also be a source of satisfaction for the acting individual.[3] In this sense, the very awareness of other people's needs can be a source of discomfort.

A person's conduct is connected with mental activity consisting of formulating and ordering goals, choosing the means to achieve them, planning and determining when the goal will be achieved. Selection of goals is based on the values attributed to these goals and understood as satisfaction stemming from their achievement. In this sense, a person pursues the goals that are most valuable to him or her. This allows us to present any human action as an attempt to replace a less satisfactory state with a more satisfactory one. The value attributed to the goal itself is not objective, but results from a subjective evaluation by particular individuals. And the pursuit of a goal means that a person identifies the means to achieve it. These means are assessed for their usefulness in achieving particular goals and in terms of the value of the goals. So the means exist only in relation to the goal they make possible. Furthermore, just like the goals, the means are subjective and depend on time and circumstances. The purposefulness of an action also means that it is based on a person's capacity for reflection, used for selecting goals and means. Thus, action is different from "unreflective reflexes".

Interpreting social reality in terms of human action means that the goals are treated as ultimate.[4] This is linked to an approach in which defining satisfaction, and therefore the goal, is a matter of individual and subjective estimation of values, different from moment to moment and from individual to individual. This means that the system of values, and thus also the goals pursued, are expressed in action. In this sense, value "is the way in which man reacts to the conditions of his environment" (Mises 2008, 96). This approach, although close to Weber's perception of goals and means as ultimate elements of human action (see Weber 1949, 52), does not deny the existence of both biological and socio-cultural factors determining human actions. It is a manifestation of an anti-naturalist approach, according to which mental phenomena cannot be entirely reduced to biophysical ones.[5] Action is not a form of an automatic reaction of the organism, but conscious management of human behaviour.

The approach to human action presented above implies that every action is necessarily rational. This rationality is based on the very purposefulness of an action and the perception of the goals as ultimate. It means identifying the goal with satisfying some need of the acting person. As a result, "[w]hen applied to the ultimate ends of action, the terms rational and irrational are inappropriate and meaningless" (Mises 2008, 18). At the same time, the selection of the means to achieve a given goal is rational. The means may be inadequate and unsuitable for achieving the chosen goal, due to the limitations of human knowledge. However, while the resulting actions may be inconsistent with the adopted goal, they still remain rational. They result from a rational attempt to achieve the adopted goal. In other words, rationality is anchored in the purposefulness of the action. In this sense, the opposite of rational behaviour is not an irrational action, but unreflective behaviour, an automatic and instinctive reaction of the organism, independent of human will.[6] This means that human action and rationality are two sides of the same coin.

Rationality perceived in this way differs from the neoclassical model of *homo economicus*.[7] This model accepts the concept of the rational man as not making mistakes in his actions. This effectively leads to identifying rationality with possessing unlimited knowledge. But the vision of man adopted by the theory of spontaneous order conceives rationality as the ability to pursue one's preferences and make choices. In particular, it allows for the existence of limitations in the knowledge of the acting individual and the influence of emotions on his or her decisions.[8] Humans make mistakes, which means that their actions do not necessarily lead to the chosen goal. However, this does not undermine the rationality of these actions. Rationality determines the formal relationship between goals and means, and does not depend on the adequacy of the knowledge possessed by the acting person. As James Buchanan pointed out in his description of the praxeological approach, "A person chooses that which he chooses, and when he so chooses, he must anticipate that the chosen course of action will yield a net increment to his satisfaction. Although he may err, we can never infer, ex post, that he acted irrationally" (Buchanan 1982, 14). Mises (2008, 14) described an individual so conceived as *homo agens* – acting being.

Methodological individualism and subjectivism

Methodological individualism

Basing the theory of spontaneous order on the category of human action means that social phenomena are perceived as an expression of the actions of individuals, thus constituting the methodological individualism of the approach: "All social phenomena (their structure and their change) are in principle explicable only in terms of individuals – their properties, goals and beliefs" (Elster 1982, 453).

At the core of such an approach lies the concept of viewing social reality on the basis of epistemological anti-naturalism (see Risjord 2014, 8–9). This concept, sometimes labelled as social anti-naturalism, proclaims the distinction between the types of cognition in the social sciences and in the natural sciences, emphasising the quality of understanding (*Verstehen*) in the former. It indicates the existence of social phenomena that cannot be explained by the scientific method, where the explanation is understood as formulating laws governing these phenomena, and the scientific method is understood as explaining things only in terms of natural phenomena.[9] Thus, it is the opposite of epistemological naturalism, which recognises the cognitive universality of scientific methods. Anti-naturalism stresses the importance of interpreting social phenomena, recognising the cognitive value of qualitative research methods. It also rejects the objectivity typical of the natural sciences.

The anti-naturalistic perspective manifests itself in the concept of action that underlies the description of social phenomena. While human action results from the meanings a given person attributes to the surrounding reality, the process of attribution cannot be adequately described in the language of the natural sciences. This is the difference between action and automatic, passive reaction to external factors. The latter, described in terms of biological or chemical processes, is deterministic. But action is based on the existence of an irreducible phenomenon consisting in the attribution of meanings to external factors, as well as to the person's own actions. This constitutes a non-deterministic – from the point of view of natural sciences – and hence authentic choice made by the acting person. This choice can be interpreted as an expression of human will.[10] This means that the environment influences the person's actions through the meanings he or she attributes to the perceived aspects of the environment.

Phenomena available to cognition are those that the actors regard as such. This indicates the impossibility of a full description of social phenomena in the terms of the natural sciences and the need to invoke the process of understanding them (*Verstehen*). This means that actions manifested in and shaped by given meanings become the subject of interpretation. Thus, the perception of social phenomena as consisting of human actions is also based on the interpretation of those actions.

The claim that the "hermeneutical horizon [...] matters" (Di Iorio 2015, 38) does not deny the role of factors external to the actor. Moreover, it does not mean the rejection of the strictly material character of the actor, understood as his or her belonging to the world described by the natural sciences, while rejecting the existence of the metaphysical aspect. However, it indicates the inability of the natural sciences to explain how the above physical factors and the physical structure of cognitive systems lead to the emergence of certain beliefs and values. Consequently, an action, as resulting from an interpretation of perceived stimuli, cannot be fully explained in the form of articulate and unambiguous rules.[11] This irreducibility also means that ideas and beliefs are the ultimate cause of human action. They are therefore the starting point for the study of social phenomena. This leads to the ontogenetic claim that an action results from the ideas entertained by the actor.

The second factor behind methodological individualism is the ontological assumption that only man is the subject of action (only man acts). In the light of the irreducibility of action, this means attributing causality only to man. So explaining social phenomena through indicating their causes forces us to name the beliefs and ideas possessed by individuals as the ultimate cause of these phenomena. It is man who constitutes the elementary unit of the social life of a community, and social structures do not exist independently from man. This also means that all social phenomena manifest themselves in human action. In other words, they are perceived in terms of actions of the individual people that constitute them. In the light of the anti-naturalistic embedding of action in the beliefs of the actor, social phenomena are construed as manifestations of the views and ideas behind the actions forming these phenomena. Consequently, institutions, conceived as sets of rules and moral norms imposing restrictions on individual behaviour, exist as beliefs functioning in the minds of people.[12] At the same time, these beliefs manifest themselves in human action. Thus, "[t]o explain social institutions and social change is to show how they arise as the result of the action and interaction of individuals" (Elster 1989, 13).

Methodological subjectivism

Basing the description of social phenomena on the category of human action affiliates the methodological individualism of the theory of spontaneous order with methodological subjectivity. The latter, conceived as recognising opinions and beliefs of human actors as facts in the social sciences, is not so much one of the methodological aspects as a key issue enabling the study of manifestations of human action and thus social phenomena as such (Horwitz 1994, 17). While social phenomena are described in terms of human action, the action itself results from the beliefs of the actor. The subjective mental states of actors are another starting point for research. As Hayek pointed out:

> In fact, most of the objects of social or human action are not "objective facts" in the special narrow sense in which this term is used by the Science and contrasted to "opinions", and they cannot at all be defined in physical terms. So far as human actions are concerned the things are what the acting people think they are.
>
> (Hayek 1955, 26–27)

This means that social reality does not exist beyond the meanings given to it by individuals:

> If the mental attitude no longer exists, society does no longer exist either. If people were not aware of each other's existence, society would not exist, even if all the same people were still in existence. [...] Thus society is an attitude in the mind of the individual which is subject to *X* changes each second. It is unstable and undermined, although it may appear constant and concrete on the surface during long periods, or made to appear this way to the social theorists.
>
> (Landheer 1952, 22)

However, this approach is far from relativistic. Society is not an atomised collection of fully independent entities. Perception of social phenomena takes place on the basis of a set of meanings and interpretative patterns common to actors forming a given social reality. This allows us to perceive society as an intersubjective construction, built by human interpretations.

The existence of a common hermeneutical perspective is based on the existence of a human species community and sharing of a given social environment by individuals. The first of these elements imposes a certain similarity on the cognitive structures and minds of individual actors. At the same time, the cognitive structure is not immutable, defined once and for all. It is not static, but dynamic, depending on the environment. The environment, by influencing people, shapes their cognitive structures. Thus, the mind is the product of a continuous process of experiencing.[13] This mind has a twofold meaning here. First, due to the similarity of its structure in various humans the influence of the common environment leads to the creation and assimilation of similar cognitive categories and interpretative patterns. Second, this means that the resulting set of common categories and patterns reflects patterns typical of a given cultural tradition. In other words, the set is assimilated in a form similar to that in which it previously existed in the minds of individuals forming the environment. As a result, both these aspects allow different actors to interpret the phenomena they perceive in a similar way. In the context of endowing human actions with meaning, the existence of intersubjective reality provides an individual with the ability to interpret actions in a way consistent with the meanings that other individuals attribute to those actions.

Consequently, subjectivism so conceived means understanding human action through being in the community. This ability is essential for the existence of social phenomena, because the convergence of meanings given to something by different people allows individuals to form expectations concerning the reaction of others to their actions, making interactions between people non-chaotic.[14] As Alfred Schütz wrote, "The prototype of all social relationship is an intersubjective connection of motives" (1960, 215). This means that the inter-subjective nature of meanings allows for mutual understanding and thus the possibility for individuals to adjust their actions, leading to the emergence of patterns of behaviour and social phenomena. It also indicates, that just like individuals creating individual social phenomena, researchers of these phenomena must rely on the same ability to understand the instances of human conduct that form these phenomena.

Importantly, the possibility – resulting from the intersubjective nature of human conduct – of anticipating the actions of other individuals and thus entering into intentional, non-random relations with them does not imply the existence of a deterministic description of the mechanism producing these actions. This view avoids reducing man to the role of an automaton reacting to external stimuli. While the description of these mechanisms belongs to the category of explicit knowledge (verbalised knowledge, "know that"), understanding human action relies on tacit knowledge ("know how").[15] In this sense, understanding intersubjectivity allows us to see social reality by means of "imposing an intellectual order on apparent chaos without inferentially reducing the status of man" (Buchanan 1982, 16).

The fact that the knowledge of meanings is shaped by a given social environment does not mean that people from outside that environment reject the possibility of knowing those meanings. Despite changes in cognitive systems that differentiate people (e.g., due to their belonging to different cultures), the very process of these changes is based on the biological similarity of the basic structures of the human mind. While people differ in their belief structure, this structure itself is based on certain common cognitive mechanisms. In this sense, the study of social phenomena by looking into the beliefs of acting individuals means that

> the individuals which compose society are guided in their actions by a classification of things or events according to a system of sense qualities and of concepts which has a common structure and which we know because we, too, are men; and that concrete knowledge which different individuals possess will differ in important respects.
>
> (Hayek 1955, 33)

This produces a non-relativistic hermeneutics, which determines the understanding of human actions and thus the possibility of studying social phenomena. The subjectivism of this approach is not the same as arbitrariness (Zanotti 2007, 116). Moreover, non-relativistic hermeneutics creates a culture-independent sphere of knowledge, which Mises describes as praxeology – the science of human action – and which concerns what is necessary in human action. The category of human action, as underlying all cognition of social phenomena, is universal (Di Iorio 2010, 193).

Founding praxeology on the community of species, and thus the common structure of minds, means that decisions are made on the basis of the same framework of rational choice for all people, regardless of the knowledge possessed by individual persons (Boettke 1998). The cultural and historical context, influencing individual ideas and beliefs, produces differences in the interpretation of needs and reality. This does not, however, deny the existence of categories and structures common to all within the human mind (Tooby and Cosmides 1992). The category of human action is something given ultimately and not subject to empirical verification, which, in Mises' view (1962), produces a priori praxeological knowledge.[16] Thus, praxeology constitutes the basic theoretical framework for the interpretation of empirical data concerning social reality.

Explaining social phenomena in terms of the intersubjective sphere of meanings brings this approach close to the perception of reality proposed by social phenomenology (phenomenological sociology). The latter combines Max Weber's individualistic and subjective sociology with Edmund Husserl's phenomenological perspective.

In sociology based on *Verstehen*, the interpretation of social phenomena involves the use of the concept of intentionality, present in every human action. Alfred Schütz, considered the father of social phenomenology, points out that if social processes are seen as an outcome of individuals' actions, the explanation of these processes must involve subjective intentions and actions of people:

> I cannot understand a social thing without reducing it to the human activity which has created it and, beyond it, without referring this human activity to the motives out of which it springs. I do not understand a tool without knowing the purpose for which it was designed, a sign or a symbol without knowing what it stands for, an institution if I am unfamiliar with its goals, a work of art if I neglect the intentions of the artist which it realizes.
>
> (Schütz 1960, 211)

Accepting the phenomenological perspective means rejecting an approach that sees social reality as objective and permanent, for social reality is a product of the activity (thinking and acting) of the people composing it. This means that it is created by objects whose meaning results from the meanings attached to human actions. This activity is based on the existence of a common environment (for the actors), which in itself is based on the sharing of cognitive patterns by individuals. Thus, reality is an intersubjective construction. As a product of the activity of individuals, it is not embedded somewhere beyond individual actors, but in the meanings contained in their minds. The world is seen not as an objectively existing reality, but as a phenomenon – it is what it appears to be. It must be observed and studied as it is experienced, thus suspending beliefs about both its reality and the reality of the actor.[17] Such a world is called the social world, or the lifeworld (*Lebenswelt*). It is a world of experiencing, formed by experiences subjectively perceived by man as the real world. This can be identified with the everyday reality of man – the totality of experience and ways of understanding it.

The interpenetration of the cognitive horizons of individual actors turns this world into an intersubjective collection of meanings, making the comprehension of social facts possible. Such a world can be identified with the everyday reality of man – the totality of experiences and ways of understanding them. As a "grand theatre of objects variously arranged in space and time relative to perceiving subjects, it is already-always there, and is the 'ground' for all shared human experience" (Husserl 1970, 142). For Edmund Husserl, the social world forms the basis for determining the truth of a shared experience, and thus its significance for epistemological analysis.[18]

In the context of methodological individualism and subjectivism, the phenomenological social world exists because of the experiencing person (the "I"). For it is the individual who gives meaning to the objects of this world through his actions, and it is for the individual actor that this world has meaning. However, this does not mean that this world is experienced individually, that is independently and in isolation, by each individual. No: as conceived by social phenomenology, the individual shares it with other people. While meaningful objects are created by individuals, this is done within the framework of interaction with others. The social world understood in this way is uniform for the people who belong to it, and experienced together. This means that it is intersubjective from the very beginning.[19] In phenomenological terms, it is the intersubjective nature of the world that determines the subjectivity of an individual's perception of it rather than the other way around. Moreover, such a world determines the intersubjective nature of individual goals. This in turn allows for their mutual adjustment in the process of interaction, providing the basis for the existence of social relations.

Between atomism and holism

Accepting an intersubjective view of the world means that perceiving social phenomena as emerging from human action does not undermine the importance of structures and institutions, and their impact on the individual. Humans always find themselves in a certain institutional context that influences their actions. This means that the discussed approach differs from atomic methodological individualism, based on the perception of people as a kind of "a collection of Robinsons Crusoes" (M. Friedman 1962, 13) – autonomous actors abstracted from their communities. The latter approach, ignoring the importance of institutions and the role of mutual relations between individuals, reduces the reality under study to an aggregation of activities of isolated individuals. Opposing this view of social phenomena, the theory of spontaneous order does not deny man his social character:

> When he [individual man] is born, he does not enter the world in general as such, but a definite environment. The innate and inherited biological qualities and all that life has worked upon him make a man what he is at any instant of his pilgrimage. [...] Inheritance and environment direct a man's actions. They suggest to him both the ends and the means. He lives not simply as man *in abstracto*; he lives as a son of his family, his race, his people, and his age; as a citizen of his country; as a member of a definite social group; as a practitioner of a certain vocation.
>
> (Mises 2008, 46)

This means that social structures, insofar as they arise from human activity, affect that activity in their turn. In other words, they are both a product and a factor of human action. Similarly, human actions both create social structures and are shaped by them. Thus, each of the two elements is partially explained by the other. In the light of the feedback between individuals and institutions, explaining social phenomena only in individualistic terms without taking into account the relationships between particular individuals is highly problematic (Hodgson 2007). Such an explanation, as well as an explanation embracing only social structures, would entail reducing one of the elements of the feedback cycle to the other. In view of the mutual influence of these elements, it would mean singling out the primary aspect of this coupling. In the context of explaining the social world through individual conduct, this would imply the existence of an initial state devoid of any social structures, of a human collective from which these structures originated. However, such a reduction is impossible without additional assumptions. This is because the concepts of man and community, as reflecting the relationship between part and whole, are interdependent (Mises 2008, 42). This means that neither of these concepts is primary in relation to the other. The interdependence of both aspects is not time-dependent. In particular, the individual is not prior to the community.

The irreducibility of social phenomena resulting from the recognition of the social nature of man turns the problem of the choice of methodological perspective into the question "Which came first – the chicken or the egg?" This means that any attempt to explain social phenomena through just one of the interdependent aspects leads to infinite regress. As Geoffrey Hodgson writes, "It is simply arbitrary to stop at one particular stage in the explanation and say 'it is all reducible to individuals' just as much as to say it is 'all social and institutional'" (Hodgson 2004, 19). Without additional assumptions, the dispute over the primacy of any of these perspectives seems futile (see Nozick 1977; Ruben 1985).

The above considerations show that the problem of justifying the methodology adopted depends on the ability to define the concept of man and social structures. As Lars Udehn notes in the context of methodological individualism, "To suggest that the 'truth' of methodological individualism is secured by the fact that social wholes are made up of individuals and their relations to one another is to beg the fundamentally important questions: 'What is an individual?' and 'What is a social relation?'" (Udehn 2001, 2). These issues raise the question of how much this approach merely points to the importance of individuals in explaining social phenomena and how much it implies focusing exclusively on individuals in the explanation.[20] For even if we accept that "all social phenomena are created, or caused, by individual human beings" (Udehn 2002, 489), it is an ontological claim (see Udehn 2001, 2; Evans 2010, 7) and as such it does not have to imply specific methodological solutions (Kincaid 1997, 4; Hodgson 2007, 214–215). However, it is no less legitimate to formulate a claim – the antithesis of ontological individualism – according to which individuals are caused by institutions. Each of the two claims points to a specific form of causal relationship between an individual and an institution, but give it opposite directions. So the two approaches are symmetrical.

The issue can be illustrated in the form of a directed bipartite graph.[21] The set of this graph vertices decomposes into two disjoint sets – the set of individuals and the set of institutions – each of which is an independent set of the graph. The directed edges of the graph correspond to the causal relations between individuals and institutions. The absence of additional assumptions concerning causal relationships makes it possible to assume that for each of the directed edges there is an edge that connects the same vertices and has the opposite direction. As a result, the two independent sets (of individuals and institutions) are indistinguishable from each other in terms of the nature of the edges which connects them. This means that the form (individualistic or holistic) of description of social phenomena is irrelevant.

If we are to assume that explanation is based on causality, the choice of methodological perspective depends on how we understand it. This, in turn, requires an answer to the above questions about the individual and social relations, as this answer determines the understanding of the concept of causality.[22]

As indicated earlier, the methodological individualism of the theory of spontaneous order is based on epistemological anti-naturalism and the ontological assumption that only humans act. The first of these assumptions defines human action as non-deterministic, which follows from the impossibility of reducing the meanings given to the observed social reality by the actor to objectively describable physical stimuli. In this approach, beliefs are seen as the ultimate cause of human actions. Together with the second assumption, which identifies the social actor with the individual, this means that social phenomena, as manifested in the actions of individuals, are constructed by the beliefs leading to these actions.

This approach means that social structures are collective concepts present in people's minds and manifested in their actions. In other words, beliefs and ideas contained in people's minds are the intension of these structures and human action is their extension.[23] This means reducing the observable social world to individual actions. However, this reduction is epistemological and as such does not imply specific methodological conclusions. They are made possible by giving the beliefs of human actors the character of ultimate causes. This means that beliefs cannot be fully explained by other causes, implying that they must be explained for each individual separately. Thus, the claim that social structures result from these beliefs is not only epistemological, but also methodological.[24]

Attributing agency only to humans means that ultimate causes refer only to their conduct. There are no irreducible institutional reasons. This makes it possible to solve the problem, which at the methodological level is posed by the interdependence of the concepts of the individual and society. The absence of other agencies apart from man suggests that individual beliefs are the only available means of explaining social phenomena, conceived as indicating their causes. This leads to a methodological attitude of individualism. Moreover, since in this view social phenomena can be fully explained in terms of individuals, this approach constitutes the so-called strong methodological individualism.[25]

Methodological individualism can be regarded as a manifestation of a certain asymmetry in the relation between the concepts of individual and structure. It results from the anti-naturalistic concept of the human actor, whose conduct is based on irreducible ideas and meanings contained in his mind, and from the identification of this actor with the individual. Consequently, it is the views contained in the human mind that give direction to social causality. It should be noted, however, that reducing social structures to individual categories does not undermine the reality of the former (see: Di Iorio 2015, 103–105). Their reality results from the fact that they manifest themselves in the actual activities of individuals. In other words, they are real because, like beliefs, they influence people's conduct. In this sense, they are a cause of human activities, although this cause does not determine them and is itself located in the individual sphere, specifically in people's minds.

This approach allows us to describe relationships between structures/institutions and the individual in the form of feedback. Under this feedback, institutions, understood as formal and informal rules, create incentives influencing decisions made by the individual. And the institutions themselves, as resulting from interaction between people, are shaped (and changed) by individual beliefs. However, while the stimuli created by institutions and individual beliefs affect each other, this coupling is not symmetrical, because while the beliefs of individuals constitute an ultimate and irreducible cause, stimuli produced by an institution are reducible to the actions that constitute it and, consequently, to the beliefs of the individuals behind these actions.

The above observations indicate that methodological individualism in the form described does not undermine the role of social structures but shows that, as they are contained in the mind, they require explanation in individual terms, and not the other way round:

> It is uncontested that in the sphere of human action social entities have real existence. Nobody ventures to deny that nations, states, municipalities, parties, religious communities, are real factors determining the course of human events. Methodological individualism, far from contesting the significance of such collective wholes, considers it as one of its main tasks to describe and to analyse their becoming and their disappearing, their changing structures, and their operation.
>
> (Mises 2008, 42)

This allows us to adopt the attitude of methodological individualism while avoiding the problem (see above) of the interdependence of the concepts of individual and structure. The reality of social structures conceived in this way means that people are social beings also at the individual level, giving meaning to particular aspects of their surrounding reality. This results in a methodological perspective that combines strong methodological individualism with the attribution of a social character to the individual and thus a rejection of the atomic version of methodological individualism. This approach can be described as social individualism (Udehn 2001, 348).

It is debatable whether a full methodological reduction is possible in practice, or whether it is only reduction "in principle". Reduction "in principle", as long as it does not undermine the reducibility of institutional terms to individual ones, indicates the presence of practical limitations in the application of strong methodological individualism.[26] This is due to the complexity of social reality, as well as the limitations of our knowledge, which compel us to use terms referring to macro-scale phenomena in the explanation. According to John Watkins (1957), these phenomena constitute some kind of half-way solutions in our striving for explanations in individual terms. As knowledge about them increases, they are reduced to individual phenomena.

If one agrees that it is impossible to achieve full knowledge, there will always be descriptions of macro-scale phenomena that cannot be reduced under a given state of knowledge. Thus, although the desire for explanation in individual terms leads to an approximation of the state of full reduction, this state is not realistically achievable. This means that the approach of strong methodological individualism is seen as an endless process. This does not, however, call into question the very existence of the desire to formulate explanations in individual terms, combining both versions of reductionism. In this sense, the position of strong methodological individualism seems to find an adequate expression in the words that "Although in modern economics, collections of individuals are sometimes treated as 'entities' for analytical purposes (examples of 'the household', 'the firm', and even occasionally 'the state' spring to mind) the ultimate unit of analysis is always the individual; more aggregative analysis must be regarded as only provisionally legitimate" (Brennan and Tullock 1982, 225).

A similar approach can be found in Ludwig Lachmann's view that social sciences should make "the world around us intelligible in terms of human action and the pursuit of plans" (1976, 261–262). This aspiration, regardless of the practical aspect of reduction, means that the use of institutional concepts should be treated as a form of mental shortcut or the effect of a temporary impossibility of explaining them. In particular, the use of institutional terms without the awareness that they are embedded in the activities of individuals produces the risk of formulating collectivist theories and anthropomorphising institutions by assigning them the ability to act. According to Schütz, this is linked to the natural human tendency to personify these concepts, which results in "treating them as if they were real persons known in indirect social experience" (Schütz 1972, 198). Along with the resulting existence of a "collective" mind, this leads to these concepts being assigned their own goals and interests, independent of the interests of individuals. Consequently, this approach is seen as not only erroneous, but also dangerous, since it contributes to the emergence of a collectivist ethic that subordinates the aims of individuals to those of the community and thus reduces man to the role of a means to achieve the aims of the community (Hayek 1944; Mises 1985, 59–61; Popper 1957).

Perceiving social phenomena in terms of the human actions that constitute them means that the understanding of social reality is based on the meaning people assign to the observed actions. For reality is not given objectively, but is a construction based on irreducible beliefs and ideas contained in people's minds. It is created from the meanings given to the perceived stimuli. This leads to the rejection of both the holistic approach and the positivist attempts to explain phenomena on the basis of objective quantities. Both institutions and statistical communities are not direct data, but result from interpretations of human actions according to a certain concept of reality. Theories that serve to study institutions and aggregates as holistic entities, but ignore these theoretical ramifications, were labelled by Karl Popper as naive:

> It completely overlooks the fact that these so-called social wholes are very largely postulates of popular social theories rather than empirical objects; and while there are, admittedly, such empirical objects as the crowd of people here assembled, it is quite untrue that names like "the middle class" stand for any such empirical groups. What they stand for is a kind of ideal object whose existence depends upon theoretical assumptions.
>
> (Popper 1962, 341)

In particular, this leads to the criticism of quantitative methods that deny causality to individuals. Instead, they seek regularities resembling scientific laws among measurable values, while ignoring the theoretical ramifications of their observations. As Hayek pointed out, "It not only leads frequently to the selection for the study of the most irrelevant aspects of the phenomena because they happen to be measurable, but also to 'measurements' and assignments of numerical values which are absolutely meaningless" (Hayek 1955, 51).

While social phenomena constitute a theoreticised picture of human activities, the activities themselves are no longer reducible. This means that it is impossible to fully describe phenomena based on human action through their physical attributes. Concepts such as "tool" or "money" must conform to the meanings given to them, and thus to the purposes for which people believe they can be used (cf. Taylor 1971). In other words, their definitions "are abstractions from all the physical attributes of the things in question and [...] must run entirely in terms of mental attitudes of men towards the things" (Hayek 1955, 27). Thus, unlike in the natural sciences,

> When we study human societies the purposes we attribute to the objects of our examination are not metaphorical but real and already meaningful to them. We are able to view them "from the inside". [...] The task of the social scientist is to find and explicate a meaning that is always already there, rather than to invent a merely metaphorical "meaning" which works in predictive tests.
>
> (Lavoie 2011, 107)

In the absence of any reference to the category of human action, institutional constructs or statistical quantities are not real causes of the observed phenomena. Thus, regularities that emerge from the description of social reality in their terms cannot be scientific laws, but present only certain historical knowledge bound to a given time and place. Therefore, the theoretical aspect of these constructs aside, the aim of the social sciences is not to empirically discover laws concerning complex social phenomena treated as wholes. Grounding these phenomena in human action means that "[t]heir task is rather [...] to constitute these wholes, to provide schemes of structural relationships which the historian can use when he has to attempt to fit together into a meaningful whole the elements which he actually finds" (Hayek 1948, 72).

The recognition of holistic terms and aggregates as conceptions does not deny their possible impact on social phenomena. They are created not only by researchers, but above all by individuals who constitute the observed phenomena and so influence the actions of these individuals. As has already been mentioned, they are not objectively observed facts, but a form of conceptualization of reality and its regularities that is commonly used in everyday life. As the beliefs of the acting individuals, they are the starting point for research. In particular, this means that aggregated quantities can influence social phenomena if the actors are aware of them. Knowing them can make an impact on people's decisions, influencing the observed phenomena that result from these actions.

The approach according to which social reality is a construct based on the beliefs of individuals also leads to a different method of explaining these phenomena than in the natural sciences. This difference is due to the fact that in natural sciences the researcher has no knowledge of the ultimate causes of a given phenomenon. In the case of social phenomena, the ultimate cause is the beliefs and motivations of the individuals behind the actions forming these phenomena. In the context of epistemological anti-naturalism, the explanation of social phenomena is based on the motives behind the actions forming these phenomena, interpreted in terms of the observed actions. This means that these beliefs are the starting point for reconstructing the social phenomenon under investigation. Due to this reconstructive nature this method is referred to as composite. Alternatively, it is also called genetic-causal (see Caldwell 2004, 21–23).

This approach is the opposite of research methods used in the natural sciences, where it is impossible to reconstruct the phenomenon by invoking its ultimate cause. In the case of the latter, observation is used to determine the components constituting a given phenomenon. Such an approach results from the possibility of isolating and experimentally reconstructing particular regularities of the observed phenomena. Thus, the process of arriving at an explanation and making predictions develops through indicating the causes described by these regularities through formulating testable hypotheses. In contrast, the concept of the ultimate cause enables the composite method to use introspection in order to identify the ultimate and necessary elements of human actions, providing an insight into the "logical structure of purposive behaviour" (Wiśniewski 2014b, 41).

Knowledge, creativity and praxeology

Knowledge

The fundamental aspect characterising human actions is the absence of the full knowledge that would guarantee people that they will achieve their intended goals. In other words, human action always takes place in a situation of uncertainty. Moreover, this uncertainty is a prerequisite for action. To reject it would imply the availability of full knowledge of future events. It would mean depriving people of the possibility of making choices about their own behaviour, and thus negating the anti-naturalistic concept of human action, reducing the individual to a kind of automaton responding to stimuli without the participation of will.

The absence of full knowledge, an effect of cognitive barriers, results from the limitations of the human cognitive apparatus. Two types of such limitations can be distinguished. The first one is due to the finite data-processing capacity of this apparatus. The second follows from an assumption of epistemological anti-naturalism, pointing to the inherent inability of the cognitive apparatus to explain a certain class of phenomena related to the acting individual in the terms of the natural sciences.[27] The first of these limitations can be described as quantitative, where the limitation concerns the quantity of detail contained in the explanation of a given problem. The second explanation involves problems that are insolvable in the absolute sense.[28]

An important feature of the knowledge possessed by man is that it is not entirely explicit, that is possible to articulate. Part of it is tacit knowledge.[29] Explicit knowledge is when an individual is able to identify parts of a given phenomenon, or to determine if they are true. Tacit knowledge is when an individual knows how to perform a given activity, but does not have explicit knowledge about it (see Polanyi 1958, 1966). It expresses a subjective and personal knowledge and skills that the individual is not able to fully articulate. It produces in a situation where "we can know more than we can tell" (Polanyi 1966, 4). As a result of this and of the cognitive limitations of man, human knowledge is dispersed and cannot be aggregated and articulated. This means that every individual has a certain amount of knowledge which, because of its subjective nature, is available only to him/her. Since part of this knowledge is tacit, it cannot be fully articulated and collected in a form that would be cognitively accessible to another individual or group of persons.

Creativity

The inability, resulting from the anti-naturalistic perspective, to reduce the beliefs contained in the human mind to physical phenomena turns them into the ultimate cause of human action. This means that the production of information by the human mind does not have its own explicable cause. This creates space for the phenomenon of human creativity, conceived as the ability to produce new information *ex nihilo* (Huerta de Soto 2010c, 25). This entails the possibility of making authentic choices. As Buchanan wrote, "I am choosing what I create; I am not merely reacting to external stimuli, at least in a sense readily amenable to prediction" (Buchanan 1982, 7).

This situation does not mean that actors have full autonomy. After all, the mind, which is a source of creativity, is co-shaped by social processes based on the intersubjective world of meanings. However, it highlights the hermeneutic overtones of the discussed approach, pointing to the role of the individual as a creative source of beliefs and actions, and thus of the phenomena constituting social reality.

In the context of human action, creativity combined with the possibility of acquiring knowledge means that goals and the means to achieve them are not given to the individual from above, but are discovered by him/her. A person has the ability to see the possibility of gaining satisfaction and take action to achieve it. This ability, referred to as entrepreneurship, is a primary aspect of human action and is sometimes even identified with it (see Huerta de Soto 2008, 63), for taking action in itself means that apart from feeling discomfort, an individual can also imagine a certain desired state. Thus, an individual has the ability to discover in things and phenomena the characteristics which enable him/her to satisfy his/her needs.

It should be stressed that while the word "entrepreneurship" may suggest a close connection with market processes, it would be wrong to identify this term exclusively with the economy.[30] The general nature of the category of human action means that the term also applies to politics. It encompasses all activities aimed at the achievement of human goals, regardless of the sphere to which they relate. This also applies to the creation and transmission of information, and coordination in the political sphere.

Human creativity is also important in the context of the uncertainty of human action, because it eliminates the possibility of predicting social phenomena in a way analogous to the natural sciences. This is due to the fact that the researched phenomena, based on the activity of creative individuals, are non-deterministic. As Gerald O'Driscoll and Mario Rizzo write,

> When an individual decides to embark upon a particular course of action, the consequences will depend, in part, on what courses of action other individuals are, or will be, choosing. A world in which there is autonomous or creative decision-making is one in which the future is not merely unknown, but unknowable. There is nothing in the present state of the world that enables us to predict the future state because the latter is underdetermined by the former.
>
> (1996, 2)

In this sense, ignorance is not a transitional state. The creative nature of human behaviour indicates that access to full knowledge, guaranteeing the possibility of anticipating future social phenomena, is asymptotically unattainable (cf. Grobler 2006, 223–225).[31] This allows us to perceive social phenomena as a dynamic processes of discovering new information by particular individuals in the course of interaction with their surrounding social and physical environment, indicating an open, indeterministic character of social processes.

Praxeology

The concept of praxeological knowledge, underlying the theory of spontaneous order, is worth examining because of two related issues. The first is the criticism of the aprioristic nature of praxeology. The second concerns the division of the social sciences resulting from the adoption of the praxeological perspective.

The anti-naturalistic character of knowledge about human action, understood as a starting point for the study of social phenomena, is connected with regarding the category of human action as universal and unquestionable. This leads to its criticism, mainly from the position of empiricism and positivism, as dogmatic and non-scientific in the sense of Popper's falsificationism. In particular, the perception of praxeological knowledge as aprioristic is denied (Lewis 2010). These accusations, however, seem to be misguided and indefensible, especially in the context of the issue of theoreticising observations and the choice of theory.[32] According to Richard Bernstein (1983), each explanation possesses both a predictive aspect, relating to the possibility of testing a given theory using empirical data, and a hermeneutical aspect. The latter is connected with the Duhem-Quine problem and the need to have a theoretical foundation for empirical observations. Bernstein stresses the view that the confrontation of a theory with data can serve to undermine both the former and the latter. Testing a theory is not done in isolation, but together with auxiliary assumptions on which the observations are based. Thus, falsification relates to a greater system of knowledge rather than a single hypothesis, and does not necessarily mean rejecting the latter. This demonstrates the importance of the tacit aspect of human judgment in choosing a particular theoretical framework. In particular, it also means that praxeology can be regarded as the hard core of research programmes as understood by Imre Lakatos (see Rizzo 1982; Koppl 2002, 25–72; Zanotti and Cachanosky 2015).[33] In this sense, praxeology serves as a filter for rejecting explanations which, although consistent with empirical data, belie the concept of human action as the ultimate cause of social phenomena. This does not negate the importance of empirical data for the study of social phenomena, but indicates the existence of a theoretical framework determining the interpretation of these data. This approach is also compatible with the non-deterministic concept of man as a creative being. Indeterminism rejects the existence of laws governing social phenomena created by human activity that could be known through observation of existing patterns and regularities in the social world, in a way analogous to the methods used in the natural sciences.

The above approach led Mises (2008, 30) to distinguish two fundamental fields of social science – praxeology and history. While the former analyses the necessary and universal categories of human action, "[t]he cognition of history refers to what is unique and individual in each event or class of events" (Mises 2008, 51). In other words, history relates to what is characteristic and distinctive for a particular event. It does not focus on human action per se, but on its effects and the ideas and beliefs behind it, bringing with it empirical knowledge about a given time and place. This knowledge is based on the concept of human action and hence, unlike in the natural sciences, it can neither corroborate nor contradict praxeological knowledge:

> The study of history always presupposes a measure of universally valid knowledge. This knowledge, which constitutes the conceptual tool of the historian, may sometimes seem platitudinous to one who considers it only superficially. But closer examination will more often reveal that it is the necessary consequence of a system of thought that embraces all human action and all social phenomena. For example, in using an expression such as "land hunger", "lack of land", or the like, one makes implicit reference to a theory that, if consistently thought through to its conclusion, leads to the law of diminishing returns, or in more general terms, the law of returns.
> (Mises 2003, 2)

The distinction between praxeological and historical knowledge also produces a categorisation of social sciences that goes beyond the traditional division into disciplines such as economics, sociology or political science. It is based on the distinction between basic categories and particular phenomena defined according to these categories.

Notes

1 This description identifies the concept of consciousness with that of purpose. Every conscious behaviour is, by definition, a pursuit of purpose. In particular, this means that the phrase "conscious pursuit of purpose" is tautological.
2 "Ultimate" due to the fact that particular goals can serve as means to achieve other goals.
3 This observation leads Mises to revise the concept of selfishness: "What a man does is always intended to increase his satisfaction. Only in this sense can we use the term 'selfishness' and show that every action is necessarily selfish. Even an action aimed directly at improving the situation of others is selfish" (Mises 2008, 243). Thus, this approach does not rule out altruistic behaviour (see Załuski 2009).
4 The issue of the validity of interpreting social phenomena in terms of human action is addressed in a later part of this chapter, in the context of the methodological individualism inherent in this approach.
5 The question of anti-naturalism is discussed in more detail later in this section.
6 The issue of human mind and thought processes is presented in Chapter 10.
7 Due to the ultimate nature of the goals, the understanding of rationality presented seems to be independent of the logical system of the human mind. In the traditional understanding of rationality the functioning of the mind is based on the rules of classical logic. However, it cannot be ruled out that the mind is described by another logic, possibly rejecting the principle of non-contradiction ($\neg(p \land \neg p)$) and thus forming a self-contradictory system. In this sense, the identification of action with rationality leads to linking the problem of the conscious – unreflective dichotomy with identifying action with the issues of the logic of the human mind. See Grygiel, 2010; Hohol, 2010.
8 For more on the incompleteness of knowledge (ignorance) of man see below.
9 "Natural phenomena" means phenomena perceived intersubjectively in direct or indirect sensory cognition.
10 As Mises wrote, "the term 'will' means nothing else than man's faculty to choose between different states of affairs, to prefer one, to set aside the other, and to behave according to the decision made in aiming at the chosen state and forsaking the other" (2008, 13).
11 The issue of methodological dualism is addressed in Chapter 10. It regards, in particular, the possibility of discovering human cognitive mechanisms, the mind-brain relationship, and the related concept of "explanation".
12 The accepted understanding of the notion of "institution" is based on the definition proposed by Douglass North (1981, 201–202). He rejects the assumption that the ability of norms (making up a given institution) to constrain the behaviour of an individual is planned.
13 The concept of the human mind and cognition is discussed in more detail in Chapter 10.
14 Non-chaoticity describes a situation in which it is possible to form partly correct predictions about future states of reality. Chaoticity is the absence of non-chaoticity. Importantly, the absence of chaoticity is not synonymous with the possibility of articulating the rules for making predictions (see Chapter 10).
15 The issue of explicit and tacit knowledge is addressed in more detail in Section "Knowledge, creativity and praxeology".

16 It should be noted that the notion of aprioricity itself can be interpreted in a different way. The cognitive status of praxeology is discussed in Section "Knowledge, creativity and praxeology".
17 This process is referred to as phenomenological reduction (*epoché*).
18 This also applies to scientific experience, including the natural sciences. The accepted concept of the social world makes physical objects and concepts related to their perception also belong to it. In this sense, there is no science independent of this world.
19 While the social world is intersubjective, the aim of the science that studies it is to objectify it. Thus, social sciences are objective "meaning-contexts of subjective meaning-contexts" (Schütz 1972, 198).
20 Different possibilities for answering this question produce a wide range of approaches within methodological individualism. They range from atomic forms of methodological individualism to approaches recognising the existence of structures independent of the individuals composing them (see Udehn 2001).
21 A directed graph is a structure consisting of a set of vertices (nodes) and a set of edges. Elements of a set of edges of a directed graph are directed edges, each of which consists of an ordered pair of vertices of this graph. An independent set of a graph is a subset of vertices where no two of these vertices form an edge of the graph. A directed bipartite graph is a directed graph whose set of vertices can be divided into two disjoint and independent sets.
22 Leaving aside the issues of non-scientific determinants, a certain justification for the methodological choice of individualism can be found in the physical (material) cohesion of the individual, making him different from social structures. Along with Ockham's razor, this provides a premise for adopting the principle of methodological individualism. It should be noted, however, that such a justification, as based on a physical perception of the world, entails a strictly naturalistic view of science itself.
23 The intension of a word or symbol is called its meaning, the inner content that constitutes its formal definition. The extension of a word or a symbol is the totality of objects to which it refers.
24 As noted earlier, the very claim that individuals form social structures is ontological.
25 The perception of social relations as ideas and meanings contained in the minds of individuals is well illustrated by the institution of money. Money is a means of exchange. Its value does not result from its ability to directly satisfy human needs, but from its ability to be exchanged for goods that satisfy those needs. This means that the existence of money is only possible thanks to the idea that makes people exchange certain goods for objects called money that are of direct use to them. As a result, by allowing things representing money to be given meaning, the idea makes it possible to acquire the good we want in exchange for a piece of paper or to change the binary sequence on the server of certain financial institutions. In a similar way, we will probably be more inclined to stop the car at the sign of a waving person if we only give this situation the meaning of road control, rather than trying to hitchhike.
26 While reduction "in principle", a form of compromising methodological individualism, contains features that link it to both weak and strong methodological individualism, it can be classified as strong methodological individualism, as it starts from the idea of explanation only at the individual level (see Udehn 2001, 348–349).

27 The identification of research agents with researched agents (because only humans can be agents) means that epistemological anti-naturalism embraces not only the researcher, but every acting individual.
28 The concepts of explanation and insolvability, as well as the issue of cognitive limitations, are discussed in detail in Chapter 10.
29 The general acceptance of the distinction between explicit and tacit knowledge is attributed to Gilbert Ryle (Ryle 1949; Huerta de Soto 2010c, 41).
30 The word entrepreneurship etymologically derives from the Latin *in prehendo* meaning "to discover", "to see", "to understand". And the word "enterprising" in a broader sense means a person who "takes something on himself", i.e., is eager to undertake tasks.
31 In particular, this means that social sciences do not create predictions, but try to make the studied phenomena comprehensible in terms of past events.
32 The epistemic status of praxeological knowledge is the subject of numerous works (see: Machlup 1955; Rothbard 1976; Caldwell 1982, 1984; Stringham and Gonzales 2009; Zanotti and Cachanosky 2015).
33 While this allows us to identify the hard core with a priori knowledge, this approach rejects the position of extreme apriorism, conceived as recognising a given field of knowledge as completely aprioristic, with no room for non-aprioristic auxiliary hypotheses (cf. Block 2003; Caplan 2003).

Bibliography

Bernstein, Richard J. 1983. *Beyond Objectivism and Relativism: Science, Hermeneutics, and Praxis*. Philadelphia: University of Pennsylvania Press.

Block, Walter. 2003. Realism: Austrian vs. Neoclassical Economics, Reply to Caplan. *Quarterly Journal of Austrian Economics* 6 (3): 63–76. doi: 10.1007/s12113-003-1024-0.

Boettke, Peter J. 1998. Rational Choice and Human Agency in Economics and Sociology: Exploring the Weber-Austrian Connection. In *Merits and Limits of Market*, edited by Herbert Giersch, 53–81. Berlin: Springer-Verlag.

Brennan, Geoffrey, and Gordon Tullock. 1982. An Economic Theory of Military Tactics: Methodological Individualism at War. *Journal of Economic Behavior and Organization* 3 (2–3): 225–242. doi: 10.1016/0167-2681(82)90019-1.

Buchanan, James M. 1982. The Domain of Subjective Economics: Between Predictive Science and Moral Philosophy. In *Method, Process and Austrian Economics: Essays in Honor of Ludwig von Mises*, edited by Israel M. Kirzner, *Method, Process and Austrian Economics: Essays in Honor of Ludwig von Mises*, 7–20. Lexington: Lexington Books.

Caldwell, Bruce. 1982. *Beyond Positivism: Economic Methodology in the Twentieth Century*. London: Routledge.

Caldwell, Bruce J. 1984. Praxeology and Its Critics: An Appraisal. *History of Political Economy* 16 (3): 363–379. doi: 10.1215/00182702-16-3-363.

Caldwell, Bruce J. 2004. *Hayek's Challenge: An Intellectual Biography of F. A. Hayek*. Chicago: The University of Chicago Press.

Caplan, Bryan. 2003. Probability and the Synthetic *A Priori*: A Reply to Block. *Quarterly Journal of Austrian Economics*, 6 (3): 77–83. doi:10.1007/s12113-003-1025-z.

Di Iorio, Francesco. 2010. The Sensory Order and the Neurophysiological Basics of Methodological Individualism. In: *The Social Science of Hayek's "The Sensory Order"*. Vol. 13 of *Advances in Austrian Economics*, edited by William N. Butos, 179–209. Bingley: Emerald Group Publishing.

Di Iorio, Francesco. 2015. *Cognitive Autonomy and Methodological Individualism: The Interpretative Foundations of Social Life*. Berlin: Springer.

Elster, Jon. 1982. Marxism, Functionalism and Game Theory: The Case for Methodological Individualism. *Theory and Society* 11 (4): 453–482. doi: 10.1007/BF00162324.

Elster Jon. 1989. *Nuts and Bolts for the Social Sciences*. Cambridge: Cambridge University Press.

Evans, Anthony J. 2010. Only Individuals Choose. In *Handbook on Contemporary Austrian Economics*, edited by Peter J. Boettke, 3–13. Cheltenham: Edward Elgar Publishing.

Friedman, Milton. 1962. *Capitalism and Freedom*. Chicago: Chicago University Press.

Grobler, Adam. 2006. *Metodologia nauk*. Kraków: Wydawnictwo Znak.

Grygiel, Wojciech. 2010. Jak uniesprzecznić sprzeczność umysłu? *Zagadnienia Filozoficzne w Nauce* 47: 70–88.

Hayek, Friedrich A. 1944. *The Road to Serfdom*. London: Routledge and Kegan Paul.

Hayek, Friedrich A. 1948. *Indivdualism and Economic Order*. Chicago: University of Chicago Press.

Hayek, Friedrich A. 1955. *The Counter Revolution of Science: Studies on the Abuse of Reason*. New York: The Free Press of Glencoe.

Hodgson, Geoffrey M. 2004. *The Evolution of Institutional Economics: Agency, Structure, and Darwinism in American Institutionalism*. London: Routledge.

Hodgson, Geoffrey M. 2007. Meanings of Methodological Individualism. *Journal of Economic Methodology* 14 (2): 211–226. doi: 10.1080/13501780701394094.

Hohol, Mateusz. 2010. Umysł: system sprzeczny, ale nie trywialny. *Zagadnienia Filozoficzne w Nauce* 47: 89–108.

Horwitz, Steven. 1994. Subjectivism. In *The Elgar Companion to Austrian Economics*, edited by Peter Boettke, 17–22. Cheltenham: Edward Elgar Publishing.

Huerta de Soto, Jesus. 2008. *The Austrian School: Market Order and Entrepreneurial Creativity*. Cheltenham: Edward Elgar Publishing.

Huerta de Soto, Jesus. 2010c. *Socialism, Economic Calculation and Entrepreneurship*. Cheltenham: Edward Elgar Publishing.

Husserl, Edmund. 1970. *The Crisis of the European Sciences and Transcendental Phenomenology: An Introduction to Phenomenological Philosophy*. Evanston: Northwestern University Press.

Kincaid, Harold. 1997. *Individualism and the Unity of Science. Essays on Reduction, Explanation, and the Special Sciences*. Lanham: Rowman and Littlefield.

Koppl, Roger. 2002. *Big Players and the Economic Theory of Expectations*. New York: Palgrave Macmillan.

Lachmann, Ludwig M. 1976. From Mises to Shackle: An Essay on Austrian Economics and the Kaleidic Society. *Journal of Economic Perspectives* 14 (1): 54–62.

Landheer, Bartholomeus. 1952. *Mind and Society: Epistemological Essays on Sociology*. Hague: Nijhoff.

Lavoie, Don. 2011. The Interpretive Dimension of Economics: Science, Hermeneutics, and Praxeology. *The Review of Austrian Economics* 24 (2): 91–128. doi: 10.1007/s11138-010-0137-x.

Lewis, Paul. 2010. Certainly Not! A Critical Realist Recasting of Ludwig von Mises's Methodology of the Social Sciences. *Journal of Economic Methodology* 17 (3): 277–299. doi: 10.1080/1350178X.2010.500503.

Machlup, Fritz. 1955. The Problem of Verification in Economics. *Southern Economic Journal* 22 (1): 1–22.

Mises, Ludwig von. 1962. *The Ultimate Foundation of Economic Science*. Princeton: D. van Nostrans.
Mises, Ludwig von. 1985. *Theory and History: An Interpretation of Social and Economic Evolution*. Auburn: Ludwig von Mises Institute.
Mises, Ludwig von. 2003. *Epistemological Problems of Economics*. Auburn: Ludwig von Mises Institute.
Mises, Ludwig von. 2008. *Human Action: A Treatise on Economics*. Auburn: Ludwig von Mises Institute.
North, Douglass C. 1981. *Structure and Change in Economic History*. New York: Norton.
Nozick, Robert. 1977. On Austrian Methodology. *Synthese*, no. 36: 353–392. doi: 10.1007/BF00486025.
Polanyi, Michael. 1958. *Personal Knowledge: Towards a Post-Critical Philosophy*. Chicago: The University of Chicago Press.
Polanyi, Michael. 1966. *The Tacit Dimension*. Chicago: The University of Chicago Press.
O'Driscoll, Gerald P., and Mario J. Rizzo. 1996. *The Economics of Time and Ignorance*. London: Routledge.
Popper, Karl. R. 1957. *The Poverty of Historicism*. London: Routledge and Kegan Paul.
Popper, Karl. R. 1962. *Conjectures and Refutations. The Growth of Scientific Knowledge*. New York: Basic Books.
Risjord, Mark. 2014. *Philosophy of Social Science: A Contemporary Introduction*. New York: Routledge.
Rizzo, Mario J. 1982. Mises and Lakatos: A Reformulation of Austrian Methodology. In *Method, Process and Austrian Economics: Essays in Honor of Ludwig von Mises*, edited by Israel M. Kirzner, 53–74. Lexington: Lexington Books.
Rothbard, Murray N. 1976. Praxeology: The Method of Austrian Economics. In *The Foundations of Modern Austrian Economics*, edited by EdwinG. Dolan, 58–77. Kansas City: Sheed and Ward.
Ruben, David-Hillel. 1985. *The Metaphysics of the Social World*. London: Routledge and Kegan Paul.
Ryle, Gilbert. 1949. *The Concept of Mind*. Chicago: The University of Chicago Press.
Schütz, Alfred. 1960. The Social World and the Theory of Social Action. *Social Research* 27 (2): 203–221.
Schütz, Alfred. 1972. *The Phenomenology of the Social World*. London: Heinemann.
Stringham, E. P., and R. Gonzales. 2009. The Role of Empirical Assumptions in Economic Analysis: On Facts and Counterfactuals in Economic Law. *Journal des Economistes et des Etudes Humaines* 15 (1): 1–11. doi: 10.2202/1145-6396.1218.
Taylor, Charles. 1971. Interpretation and the Sciences of Man. *The Review of Metaphysics*, 20 (1): 3–51.
Tooby, John, and Leda Cosmides. 1992. The Psychological Foundations of Culture. In *The Adapted Mind: Evolutionary Psychology and Generation of Culture*, edited by Jerome H. Barkow, Leda Cosmides, and John Tooby, 19–136. New York: Oxford University Press.
Udehn, Lars. 2001. *Methodological Individualism. Background, History and Meaning*. London: Routledge.
Udehn Lars. 2002. The Changing Face of Methodological Individualism. *Annual Review of Sociology* 28: 479–507. doi: 10.1146/annurev.soc.28.110601.140938.
Watkins, J. W. N. 1957. Historical Explanation in the Social Sciences. *British Journal for the Philosophy of Science* 8 (30): 104–117. doi: 10.1093/bjps/VIII.30.104.

Weber, Max. 1949. *The Methodology of Social Sciences*. Translated and edited by Edward A. Shils, and Henry A. Finch. Glencoe: Free Press.

Wiśniewski, Jakub Bożydar. 2014b. The Methodology of the Austrian School of Economics: The Present State of Knowledge. *Ekonomia – Wroclaw Economic Review* 20 (1): 39–54.

Załuski, Wojciech. 2009. *Ewolucyjna filozofia prawa*. Kraków: Wolters Kluwer Polska.

Zanotti, Gabriel J. 2007. Intersubjectivity, Subjectivism, Social Sciences, and the Austrian School of Economics. *Journal of Markets and Morality*, 10 (1): 115–141.

Zanotti, Gabriel J., Cachanosky, N. 2015. Implications of Machlup's Interpretation of Mises's Epistemology. *Journal of the History of Economic Thought* 37 (1): 111–138.

4 Spontaneous order

The idea of spontaneous order emerged from the observation that some social phenomena did not arise intentionally, but were an unintended effect of individual people's actions.[1] As Menger wrote already in the 19th century:

> Language, religion, law, even the state itself, and to mention a few economic social phenomena, the phenomena of markets, of competition, of money, and numerous other social structures are already met with in epochs of history where we cannot properly speak of purposeful activity of the community as such directed at establishing them.
>
> (1985, 146)

Importantly, the very fact of the existence of unforeseen effects of human activity, which is a natural implication of human cognitive limitations, was not seen as problematic here. The key to analysing unplanned phenomena was the observation that these phenomena were often characterised by a certain order. This meant that they could lead to the emergence of institutions commonly perceived as not only beneficial, but also crucial for the social order. The order itself should be understood here as

> a state of affairs in which a multiplicity of elements of various kinds are so related to each other that we may learn from our acquaintance with some spatial or temporal part of the whole to form correct expectations concerning the rest, or at least expectations which have a good chance of proving correct.
>
> (Hayek 1973, 36)

This means that order is a form of regularity, which determines the possibility of making correct predictions. Without the possibility of observing regularity, phenomena would be perceived as chaotic. Thus, the existence of order, both in the physical and the social world, is the basis of human action.[2]

In the case of the social world, where phenomena arise from human action, the existence of order implies the ability of individuals to create expectations about the behaviour of others in the context of their own actions. In other words, order manifests itself in the convergence of intentions and expectations of interacting people. Thus, the ability to create order can be understood as the ability to coordinate the actions of different individuals.

The existence of unplanned effects of human actions makes it possible to distinguish two types of order analysed by social sciences: planned order and spontaneous order.[3] Planned order arises from actions taken to create it, from a conscious and deliberately executed plan. Spontaneous order, also referred to as emergent or unplanned, is an order arising from unplanned processes. It does not emerge as an intentional result of a conscious decision or collective agreement. This idea is perfectly reflected in the words of Ferguson, who described it as "the result of human action, but not the execution of any human design" (Ferguson 1966, 122). This means that, unlike the planned form, spontaneous order cannot be traced back to a goal behind its creation. The only identifiable goals are those of the individuals producing this order. In this sense, the function of such an order can be identified, conceived as mechanisms enabling individuals to achieve their individual goals. As for the planned order, the existence of a goal means that it is possible to identify the decision-making structure responsible for the deliberate creation of this order. Such a governing body is absent in spontaneous order. This means that, as Polanyi (1951) notes, the spontaneous – planned distinction coincides with the distinction between mono- and polycentric orders. Monocentric order, with a central decision-making body, corresponds to planned order. Polycentric order, with no hierarchical decision-making structure, can be described as one in which "actions are determined by the relation and mutual adjustment to each other of the elements of which it consists" (Hayek 1967, 73).

The above arguments make it possible to describe the theory of spontaneous order as a research approach trying to explain the phenomenon of spontaneous order. The very possibility of examining such phenomena results from the methodological perspective adopted. While social individualism strives to explain things in terms of individuals, the key observation of the theory is that individuals possess incomplete and subjective knowledge. The existence of epistemic human limitations implies the absence of full knowledge about the effects of human actions. And this creates space for phenomena that result from human actions, but are not foreseen by the acting individuals. This means the rejection of the perception of social phenomena as planned in principle. This makes it possible to analyse and explain a number of institutions and social phenomena functioning in our reality that are not the result of individual decisions or a collective consensus aimed at producing them.

In the light of the above, the problem underlying the theory of spontaneous order may be presented in the form of the following question: How do the actions of individuals – pursuing their own particular goals and equipped with limited knowledge – lead to the emergence and enable the functioning of often complex and unplanned social structures?

The methodological perspective adopted by the theory of spontaneous order not only suggests the way in which social phenomena are explained, but is also crucial for the perception of the role of the social sciences. The resulting description of social phenomena in terms of their intended or unintended character leads to a constraining of research to the latter only. This follows from the fact that the knowledge about a phenomenon which is an intentional result of human actions is given. Otherwise, it could not be used for constructing this phenomenon. While this phenomenon may become the subject of psychological research, its explanation within the framework of the social sciences no longer makes sense. This means that the approach adopted sees explaining unplanned social phenomena as an essential role of the social sciences in general. As Hayek wrote, these sciences "are concerned with man's actions, and their aim is to explain the unintended or undesigned results of the actions of many men" (Hayek 1955, 25).

It is also worth noting that perceiving social phenomena as unintended effects of individuals' actions does not rule out an intersubjective view of the world. The existence of a common platform of communication allows people to create expectations regarding the reaction of others to their conduct. However, the very possibility of forming these expectations does not preclude the fact of the cognitive limitations of man, and thus human creativity. As a result, it does not imply access to full knowledge about the effects of interaction between individuals themselves acting in accordance with these expectations. Hence, it does not rule out the existence of unplanned institutions and social structures.

The starting point for the theory of spontaneous order is the question of the availability of the knowledge possessed by people. It is knowledge that determines human actions and hence the social phenomena resulting from them. This means that the creation of expectations about the conduct of others, and thus the possibility of coordination between individuals, is based on access to the knowledge possessed by particular individuals. This allows the theory of spontaneous order to present the emergence and functioning of unplanned institutions and structures as a result of the process of adaptation of individuals to local circumstances. All people, according to their subjective knowledge and in the course of their activities and interactions with others, discover and create new information that allows them to find new and more effective ways to achieve their goals.[4] But this information, as it influences a person's actions, can be discovered by other individuals, who can use it to achieve their own goals. So this leads to people adjusting their behaviour in the course of interactions between them. Importantly, an individual does not need to have all the knowledge about how the system works to achieve his or her goal, but uses the actions and knowledge of others.

The free market is an example of an order based on the process of adaptation to the behaviour of others (see Hayek 1945; Hayek 1976, 107–132; Huerta de Soto 2010c, 15–48). Within the market, individual actors create and communicate their own needs through their actions. This allows other participants to discover the resulting opportunity to make a profit, suggesting new possibilities of using their resources.[5] The pricing system plays a special role in the exchange of such information. It provides the actors with knowledge about the valuation of a given good, allowing them to see possible opportunities for profit without having to possess full knowledge about the valuation process and the resulting changes in individual needs.

The theory of spontaneous order also implies the existence of restrictions on the deliberate creation of social structures. The coordination of all human activities in a planned manner requires collecting all the necessary knowledge within some central decision-making body. It implies the necessity to access the knowledge possessed by particular individuals. But this knowledge is dispersed and this, combined with the epistemic limitations of man, means that it cannot be fully aggregated. Limitations in the deliberate shaping of social reality are particularly evident in the context of the authentic creativity of man and the tacit nature of part of his knowledge. All in all, the knowledge on the basis of which individuals act themselves cannot be fully aggregated and expressed. This means that no decision-making body is able to present a description of deterministic phenomena resulting from human actions, and thus to freely construct relatively uncomplicated structures.[6] But in the case of spontaneous order, the coordination process takes place not as a result of an aggregation of knowledge, but through individuals using the knowledge and the actions of others to achieve their own goals. This allows for the use of the dispersed knowledge possessed by individuals and the emergence of social structures with a complexity far exceeding human cognitive faculties. This means that all complex social structures must emerge spontaneously. Society in the broadest sense of the term, described by Hayek "the extended order", is the result of spontaneous processes and is based on

> a great framework of institutions and traditions – economic, legal, and moral – into which we fit ourselves by obeying certain rules of conduct that we never made, and which we have never understood in the sense in which we understand how the things that we manufacture function.
>
> (Hayek 1988, 14)

Interactions between individuals and the resulting ability to use information held by others for one's own purposes means that one does not need to have all the knowledge to achieve those purposes. Thanks to using the knowledge of others, people are able to achieve more. Since they do not need to reproduce the knowledge possessed by others, they can channel their mental faculties into discovering new information (see Huerta de Soto 2010c, 33–34). This process results in the division of knowledge, which is a generalisation of the division of labour observed in economics. This does not deny the usefulness of the knowledge possessed by the human actor, but indicates its significant limitations. At the same time, it underlines the importance of this knowledge for other members of society. In this sense, the increase in the human ability to take various actions and achieve more goals must be based on the knowledge that individual people do not possess (Hayek 1973, 15). This approach is well reflected in Alfred Whitehead's claim:

> It is a profoundly erroneous truism, repeated by all copy books and by eminent people when they are making speeches, that we should cultivate the habit of thinking of what we are doing. The precise opposite is the case. Civilization advances by extending the number of important operations which we can perform without thinking about them. Operations of thought are like cavalry charges in a battle – they are strictly limited in number, they require fresh horses, and must only be made at decisive moments.
>
> (1911, 61)

This underlines the importance of institutions which, as sets of rules and norms, reduce the amount of information needed to make useful predictions about human actions, helping mutual adaptation.

The above analyses indicate that the fact of the emergence of social structures does not in itself mean that man "must also be able to alter them at will so as to satisfy his desires or wishes" (Hayek 1978, 3). Consequently, the theory of spontaneous order rejects the perspective in which the institutions composing society are, or should be, products of a deliberate plan. Criticism of such an approach, described as constructivist rationalism, concerns the underlying issue of recognising the availability of the knowledge needed for governing bodies to coordinate human interaction.[7] Such an assumption is not only unrealistic, but also denies the existence of any problem requiring clarification. It suggests that all the knowledge needed to shape social phenomena is given in advance. However, as pointed out earlier, such phenomena are not the subject of study for the social sciences, because as Hayek wrote, "To assume all the knowledge to be given to a single mind [...] is to assume the problem away and to disregard everything that is important and significant in the real world" (Hayek 1945, 530). This means that the challenge to the social sciences is not the phenomena resulting from decisions made on the basis of full knowledge needed to achieve a goal, but actions based on knowledge that is not given to anyone in its entirety.[8]

The causes of perceiving social order as intentionally shaped by man can be found in the human tendency to identify order with purposefulness, and hence with the existence of an acting being to whom the intentional creation of a given order can be attributed. According to Hayek (1973, 10), from the epistemological point of view perceiving institutions as intentional human creations derives from the Cartesian concept of reason, seen as the faculty of deductive reasoning on the basis of irrefutable assumptions. This leads to the necessity of regarding all useful solutions and social structures as deliberate works of human reason (Hayek 1973, 9–10). This is closely related to the concept of the fatal conceit of reason, understood as the belief in the ability of the mind to form order at will (Hayek 1988). Barry (1982, 7) points at the success of science, based on the possibility of constructing an experiment and creating predictions, as an important causative factor here. The resulting attractiveness of scientific methods leads to the identification of all desirable outcomes with the possibility of controlling the processes producing them: "It is these methods which have an irresistible appeal to that hubris in man which associates the benefits of civilization not with spontaneous orderings but with conscious direction towards preconceived ends" (Barry 1982, 7). All of this favours assigning intentionality to observed regularities. This belief is reinforced by the difficulty with grasping the phenomenon of spontaneous order. As Ulrich Witt points out, "spontaneous order in the interactions of the members of society is something to which everyone contributes, from which everyone benefits, which everyone normally takes for granted, but which individuals rarely understand" (Witt 1994, 179).

It is worth noting that the dispersion of knowledge not only makes its aggregation problematic, but also has normative implications. Regulations introduced by a governing body, influencing the actions of individuals, affect the possibility of transmitting information between them and thus creating new information. This gives rise to fears that attempts to interfere in the social order through regulations based on constructivist rationalism will not only fail to achieve the planned result, but also disrupt the bottom-up processes of mutual adaptation of individuals, limiting their freedom to choose their own goals. The theory of spontaneous order points especially to the negative effects, resulting from coercion, of attempts to interfere with and control the free market (Mises 1990a, 1998; Ikeda 1997; Rothbard 2009; Huerta de Soto 2010c). They tamper with the otherwise free exchange of information between individuals and thus undermine mutual coordination. The resulting, crucial importance of individual freedom for the coordination of human actions makes this theory strongly related to liberalism.[9]

Notes

1 See Part I.
2 However, it should be noted that there is a significant difference between predictions in the two worlds, resulting from the creative nature of human action. Limitations in adequately predicting human behaviour produces expectations regarding actions of others based on introspection as giving "insight from within".
3 Perceiving the category of purposefulness as belonging only to the social world means that the concept of a spontaneous phenomenon can also be applied to phenomena from the physical world. In the following, the concept of order is analysed only in the context of social phenomena.
4 In this approach, the term "more effective" refers to the subjective assessment of a given measure by the acting person.
5 For example, information about the demand for a service may induce entrepreneurs to change the way they use their resources and employ them to offer this service.
6 The issue of understanding the concepts of simplicity and complexity in relation to social structures is presented in Section "Hayek's concept of the cognitive system" of Chapter 10.
7 Hayek pointed to the affinity between the idea of constructivist rationalism and Popperian naive rationalism (Hayek 1973, 5).
8 In the area of market phenomena, this leads to criticism of the concept of perfect competition, describing a situation in which all market participants possess full

knowledge. This means that there are no errors in the market, and thus no price corrections. This prevents any rivalry between the actors. So the existence of perfect competition rules out any competition (Barry 1985, 139).

9 While the theory of spontaneous order can be linked to liberal thought, the issue of liberalism itself is not a subject of this book and is not broadly presented in it. It is addressed mainly in the context of the reflection on the axiological aspect of political science presented in Section "The theory of spontaneous order in relation to the epistemic system of contemporary political science" of Chapter 5.

Bibliography

Barry, Norman P. 1982. The Tradition of Spontaneous Order, Literature of Liberty. *A Review of Contemporary Liberal Thought* 5 (2): 7–58.

Barry, Norman P. 1985. In Defense of the Invisible Hand. *Cato Journal* 5 (1): 133–148.

Ferguson, Adam. 1966. *An Essay on the History of Civil Society*. Edinburgh: Edinburgh University Press.

Hayek, Friedrich A. 1945. The Use of Knowledge in Society. *The American Economic Review* 35 (4): 519–530.

Hayek, Friedrich A. 1955. *The Counter Revolution of Science: Studies on the Abuse of Reason*. New York: The Free Press of Glencoe.

Hayek, Friedrich A. 1967. *Studies in Philosophy, Politics and Economics*. London: Routledge and Kegan Paul.

Hayek, Friedrich A. 1973. *Rules and Order*, Vol. 1 *of Law, Legislation and Liberty*. London: Routledge.

Hayek, Friedrich A. 1976. *The Mirage of Social Justice*. Vol. 2 *of Law, Legislation and Liberty*. London: Routledge.

Hayek, Friedrich A. 1978. *New Studies in Philosophy, Politics, Economics, and the History of Ideas*. Chicago: University of Chicago Press.

Hayek, Friedrich A. 1988. *The Fatal Conceit: The Errors of Socialism*. London: Routledge.

Huerta de Soto, Jesus. 2010c. *Socialism, Economic Calculation and Entrepreneurship*. Cheltenham: Edward Elgar Publishing.

Ikeda, Sanford. 1997. *Dynamics of the Mixed Economy: Toward a Theory of Interventionism*. London: Routledge.

Menger, Carl. 1985. *Investigations into the Method of the Social Sciences with Special Reference to Economics*. New York: New York University Press.

Mises, Ludwig von. 1990a. *Economic Calculation in the Socialist Commonwealth*. Auburn: Ludwig von Mises Institute.

Mises, Ludwig von. 1998. *Interventionism: An Economic Analysis*. Irvington-on-Hudson: Foundation for Economic Education.

Polanyi, Michael. 1951. *The Logic of Liberty*. London: Routledge and Kegan Paul.

Rothbard, Murray N. 2009. *Man, Economy, and State with Power and Market*. Auburn: Ludwig von Mises Institute.

Whitehead, Alfred North. 1911. *An Introduction to Mathematics*. London: Williams and Northgate.

Witt, Ulrich. 1994. The Theory of Societal Evolution: Hayek's Unfinished Legacy. In *Hayek, Co-ordination and Evolution: Hayek, Co-ordination and Evolution: His Legacy in Philosophy, Politics, Economics and the History of Ideas*, edited by Jack Birner, and Rudy van Zijp, 178–189. London, New York: Routledge.

Part III

Application of the theory of spontaneous order in political science

5 Political science and the theory of spontaneous order

Pluralism in political science

The basic issue from which it is worthwhile to proceed in considering the possibility of applying the spontaneous order theory to the study of political phenomena is the question of the very status of political phenomena, and thus of the criteria for ascribing various research perspectives to political sciences as dealing with the study of the political world. The pluralistic and multi-paradigmatic nature of political science is highlighted here (Krauz-Mozer and Ścigaj 2013). This state of affairs is based on the multifaceted and syndromatic nature of the subject of political science. The first of these features follows from the aspect-based character of cognition (Karwat 2006), for reality is not available in all its complexity, but is always the result of the perspective adopted by the conditioned agent at a given moment in time (Ścigaj 2010, 33). Cognitive limitations are also manifested in the unpredictability of individual and group behaviour. Consequently, social phenomena undergo constant change in a way that makes it impossible to define them definitively. This constitutes the syndromaticity of political reality, understood as "a tangle of heterogeneous phenomena of different origin, conditionality, varied form and different formal affiliation to different fields (technology, economy, symbolic culture), irreducible to one dimension (e.g. economic, legal or religious), and forming a unique quality" (Karwat 2009, 175). This means that the subject of research has an incidental and historically variable character, escaping attempts to place it in a rigid formal framework and imposing restrictions on the scope of validity of the theories formulated.

The absence of universal and immanent features of politicality makes it impossible to extract a homogeneous class of political phenomena (Blok 2009, 55–86).[1] The subject of the study of political science is conventional and historically variable (Karwat 2010, 78). Hence, it is problematic to try to delineate the discipline of political science by precisely defining the subject of research (see Woleński 1982, 4). Social phenomena, although differently perceived, defined and studied by individual research approaches, constitute certain wholes. In the context of the dynamic and contextual nature of political phenomena, this makes it necessary to take into account the achievements of various areas of social sciences in one's research, since what is political is "always embedded in the network of relations with other elements of social life" (Rosicki and Szewczak 2012, 49). Attempting to extract purely political phenomena would impoverish our knowledge about politics.

As with the case of the subject of research, the political sciences are also not methodologically distinct. The choice of methods depends on the subject and objectives of research rather than directly on the discipline itself. The subject of study, as has already been noted, is not unique to political science, and this means that methods may be different too. This results also from the instrumental nature of the methodology, which means that the realms of particular disciplines in the social sciences are determined not by the subject or methodology of the research, but by questions asked within these disciplines (Krauz-Mozer 2005, 14–15). Both the subject and methods form a common legacy of social sciences (Karwat 2010, 67). As Barbara Krauz-Mozer points out (1992, 11–12), all social sciences have just one subject, that is, man with his products and the society he creates. And so the methodological and subject specificity can only be attributed to such broad areas of science as humanities, for which the subject of research is the society and man understood as a social being (Klementewicz 2004, 236).

It is worth noting at this point that the existence of limitations in defining the subject and methods of political science led Tadeusz Klementewicz to suggest (1991) that two approaches to it should be distinguished: traditional political science, which perceives this discipline in terms of the subject of research, and theoretical political science, identified on the basis of relations with other sciences. An example of this is the proposal to look at the discipline in terms of its boundaries (Ścigaj 2010). This approach, based on the dialectic of identity, consists in identifying the science of politics through a comparison with what belongs to other disciplines. Thus, it focuses not on what is common to political scientists, but on what distinguishes them from others, and it is intended to find points of contact between political sciences and these disciplines.

The multitude of approaches within political sciences is a strong premise for the pluralistic nature of these sciences (Beyme 2006). This idea, understood as a position that allows different currents and research perspectives to coexist, rules out looking at political phenomena in a one-sided way. The approach to the reality under study should be interdisciplinary, capable of capturing the polymorphous nature of the examined phenomena. This allows us to distinguish two basic dimensions of pluralism so conceived (Krauz-Mozer and Ścigaj 2013, 17–18). The normative aspect indicates a multiplicity of approaches and methods that cannot be reduced to each other. The pragmatic aspect points to the instrumental nature of the methods, which should be seen as tools allowing for more comprehensive research. Such pluralism, however, must be based on the awareness of both research objectives and different methodological approaches, together with the philosophical assumptions behind them (Krauz-Mozer 2009). Hence, political science identity-building can be seen as a process of continuous dialogue with other disciplines.

The pluralism of political science expands the range of phenomena regarded as political and allows it to go beyond the traditional subjects of research, such as the state and the political system (Antoszewski and Herbut 2003, 327). The multi-faceted and syndromatic nature of social phenomena makes political science open to new issues and problems, previously examined only by other disciplines. Consequently, the concept of politics began to embrace not only the issue of public authority, but also the more general phenomenon of exerting influence (Dahl and Stinebrickner 2002, 24). This is connected with the belief in the prevalence of political phenomena as determining many spheres of social life (Nocoń 2010, 57–58). It also applies to politics, conceived in its traditional dimension as connected with the state.[2] As Anthony Giddens argues (1987, 25–33), since society is a formation defined both territorially and politically, we must understand it in terms of the concept of the nation-state. At the same time, the social order is closely linked to the political order, and Giddens questions the possibility of studying society in isolation from the state and the political system.

It is worth noting at this point that the idea of pluralism is present not only within the political sciences, but seems to characterise all areas of social science. A strong indication of this is the already mentioned fact that all social sciences have a common research subject, which is man, along with the society he creates, as well as this subject being syndromatic and multi-faceted. This leads to the absence of a clearly dominant theory, reinforcing the idea of pluralism within all social disciplines (Beyme 2006, 38). The coexistence of different approaches and schools can be observed especially in economics. This diversity and the resulting disputes within this discipline are well reflected in the words of Lionel Robbins: "We all talk about the same things, but we have not yet agreed what it is we are talking about" (Robbins 1932, 1).

The pluralism of political science makes it possible to study phenomena from different sides. This creates the possibility of obtaining both a detailed knowledge about particular aspects of these phenomena and complementary ways of doing so. Consequently, the multidimensional nature of research allows for a more precise knowledge of the research space and the creation of new research methods. This is an important advantage of political science, determining its development potential (Ścigaj 2010, 49). Moreover, as noted by Piotr Borowiec (2014, 29), it enables the realisation of the postulate of Lakatos, who stated that the development of scientific knowledge should be driven by constructive criticism formulated by competitive research programmes. The pluralism of political science also emphasises the key importance of moving between various sciences when doing political research (Karwat 2009, 188). In the pursuit of knowledge, "substantive problems are best dealt with not by utilizing one method or confining the discourse to one field, but trying to pierce the heavy curtains of instituted and institutionalized boundaries and by drawing upon as wide a range of resources as are available" (Goody 1986, vii).

While pluralism seems to be an important advantage of political science, the use of different research perspectives demands consistency: "In research, what matters is theoretical knowledge; who provides it and under what disciplinary name is of little importance if it is used consistently from the logical and semantic point of view" (Klementewicz 2013, 38). This indicates the crucial importance of being aware of the assumptions behind these perspectives and maintaining their conceptual consistency. The absence of such an awareness may lead to the error of eclecticism, consisting in losing a coherent and comprehensive picture of reality due to the introduction of concepts and claims of other theories (Klementewicz 2013, 41).

While logical and semantic cohesion is a necessary condition for the use of different research perspectives, these perspectives are not reducible to one another. This means that it is impossible to rank these approaches according to the value attributed to them, as they are formulated within different visions of science. Defining possible criteria for such a ranking would require invoking one of these visions. However, this does not mean sliding into relativism as understood by Paul Feyerabend (1993). Particular research perspectives may differ in their legitimacy, determining the degree of corroboration or the explanatory power of a given approach.

The criteria proposed by Klementewicz (2013, 39–40) seem interesting in this respect. He indicates that the assessment of particular concepts should be made with reference to their cognitive qualities, the rapid rhythm of cultural evolution, and the importance of human awareness and limited determination of social processes. As for the first factor, the greater value should be attributed to concepts that are characterised by a relational-structural approach to social phenomena and processes. In the case of the second factor, these are process-dynamic approaches, opposing the ahistorical perception of social reality. And human consciousness implies the need to take into account the creativity of the human person, resulting in the occurrence of phenomena contrary to expectations. Such an anthropoclastic perspective is opposed especially to the approaches of psychologism, structuralism and reification.

The above considerations, especially about the pluralism of political science, seem to provide a fertile ground for using the theory of spontaneous order to study political phenomena. This is in line with the multi-pronged nature of political research, which, after all, determines its development. The very possibility of implementing the theory of spontaneous order in political science also stems from its being founded on the concept of human action. Economics, the area of the most developed use of the spontaneous order approach, essentially examines the sphere of interactions between individuals, seeing these interactions as an exchange process based on the institution of private property (see Rothbard 2009, 162). In its traditional sense, the science of politics examines phenomena related to the functioning of state institutions, and in a broader sense, the notion of politics is extended to relations of influence (Blok and Kołodziejczak 2015, 23–24). This means that it encompasses phenomena going beyond the institutional framework of the free market. However, in all these cases, the studied phenomena are the result of human action, which is always characterised by the absence of full knowledge, guaranteeing that the actor achieves only the intended results. This means that the sphere of phenomena examined by the theory of spontaneous order embraces all areas of deliberate human activity, which also means political phenomena.

The theory of spontaneous order also benefits from its inclusion in the criteria of legitimacy mentioned earlier. The theory is based on social individualism, an approach that, by bringing institutions to the level of individual beliefs, promotes a relational and structural view of social phenomena. Moreover, the fact that human conduct operates in the conditions of a permanent deficit of knowledge leads to the rejection of the purposeful idea of history. Describing reality in terms of human action and the accompanying attribution of causality to man means that the theory of spontaneous order is procedural. One expression of this is the assumption of epistemological anti-naturalism, which, by seeing human action as non-deterministic, allows us to perceive man as a conscious being capable of free behaviour. Therefore, we can call the theory of spontaneous order "anthropoactivist".

It is also worth noting that under the theory of spontaneous order, the driving force of social phenomena is human action. In the context of the syndromatic nature of politics, and thus the difficulty in precisely determining what kind of action is political (see Dahl 2004), this seems to be an additional advantage.[3] In this sense, as well as in the light of our earlier analyses, particular disciplines of the social sciences are not delineated by their subject matter. They all study conscious human behaviour. At the same time, the non-deterministic nature of human action makes it impossible to assign a phenomenon to a given discipline on the basis of the external stimuli affecting the actor. What determines the nature of a phenomenon are the meanings attributed to the actions that create it. Alternatively, instead of human action one can speak about its motives, as these issues are closely related (motives manifest themselves only in action). This means that defining the boundaries of a scientific discipline on the basis of the meanings attributed to human actions seems to be an important advantage of the theory of spontaneous order – for, in principle, it does not rule out the possibility of exploring newly identified areas of politics with the use of this theory.

The theory of spontaneous order in relation to the epistemic system of contemporary political science

The absence of a unique subject or method in political science leads to the absence of ultimate and immutable boundaries in this discipline. However, it is possible to pinpoint certain aspects common to contemporary research approaches. According to Klementewicz (2014, 275), these constitute the epistemic system of contemporary political science. Examining the theory of spontaneous order in the light of this system will tell us how the new research perspective relates to the fundamental view of the world that is dominant in political science. Indirectly, it will also show us the research areas where, by solving existing problems and identifying new ones, the theory seems to provide a new, valuable perspective on political phenomena.

Klementewicz (2014, 275–279) identifies four fundamental groups of premises which are common to various paradigms of researching the world of politics and testify to the existence of the epistemic system: anthropological, sociological, methodological and axiological. The first group relates to the issue of the uniqueness of human behaviour. It includes beliefs about the species-specific nature of man understood as a social being, embedded in Darwin's theory of evolution. In terms of the cognitive apparatus and human mental dispositions, these assumptions indicate the existence of an introspective and reflexive consciousness, allowing for the objectification of thinking and will. This means that man has the ability to intentionally influence the course of his mental activity. He is equipped with the faculty of creative inventiveness, consisting in the "imaginative/intellectual creation of goal models" (Wierciński 1997, 31). Along with the possibility of managing and ranking these models, this produces the faculty of volitional behaviour. In the context of the social nature of man, these premises define his species-specific feature of satisfying his needs through collective action, in which there is a division of roles. In this perspective, people also have the ability to influence their natural environment by using tools. This influence makes it possible to create a space of collective existence, in which man becomes subjected to cultural evolution.

The second group is made up of sociological premises, illustrating two general concepts of society in political science. The first is the organic-systemic model of society, a loose adaptation of structural functionalism, ecology and cybernetics. In this concept, maintaining the social structure is possible thanks to the functions performed by four types of institutions: economic, political, established patterns of behaviour and institutions of spiritual culture along with various forms of ties and traditions. They are respectively responsible for: ensuring that people's biogenic needs are met, coordinating their actions, consolidating their patterns of behaviour, and integrating and creating a common identity. The second model is the morphogenetic (process) approach, perceiving society as a dynamic process, combining exploitation of the ecosystem by humans with the development of the human awareness of social reality.

The methodological premises are supplemented with the concept of cognitive rationality, defining the objectives and criteria of scientific cognition. The aim of political science is to explain individual and collective behaviours observed in political life, using scientific laws. These laws, allowing us to make predictions, are empirically testable and falsifiable.

The fourth group is axiological (ideological) premises, the basic one being the domination of liberal concepts of freedom and equality among political scientists. It manifests itself in the universality of the ideology of liberal democracy. It is inextricably tied to the concept of the legal state, understood as the primacy of law in the political system.[4] This means that the state must establish legal norms within these principles and in accordance with the liberal concept of individual freedom. The law perceived in this way constitutes an impersonal public authority, which is an expression of the public good and a warrant of individual rights. Klementewicz (2014, 179) also draws attention to the Eurocentrism still prevailing among Western political scientists, contrasted by him with a polycentric approach, which is closer to historical truth.

Analysing the theory of spontaneous order in the light of the presented epistemic system of political science allows us to notice that this theory is compatible with two of the three anthropological premises of this system. In particular, it corresponds to the concept of the human mind proposed by the system. The idea of human action underlying the theory creates the possibility of choosing goals and means to achieve them on the basis of expected results. This means that humans can intentionally influence their own behaviour. They also have the ability to create and rank models of reality for achieving these goals, as well as to simulate actions using these models. The theory of spontaneous order also fits into the perception of man as a social being. This results directly from the approach of social individualism, which rejects the atomistic methodological individualism. In addition, by emphasising the cognitive limitations of man it points to the importance of the transmission of information between individuals for the possibility of using dispersed knowledge and, consequently, for the mutual coordination of activities and the creation of social order.

However, the premise emphasised by Klementewicz (2014, 276) concerning the role of tools as a non-organic influence of man on the environment raises some doubts. While it is indisputable that the cultural evolution of man is connected with the spread of the use of tools and the development of technology, naming the ability to use tools as one of the key characteristics constituting the uniqueness of human behaviour seems unjustified. According to the hypothesis formulated by Michael Tomasello (1999), the driving force of cultural evolution is the unique faculty of social cognition created by man during his biological evolution. Social cognition makes the process of cultural evolution cumulative, due to what he calls the ratchet effect:

> The process of cumulative cultural evolution requires not only creative invention but also, and just as importantly, faithful social transmission that can work as a ratchet to prevent slippage backward – so that the newly invented artifact or practice preserves its new and improved form at least somewhat faithfully until a further modification or improvement comes along.
>
> (Tomasello 1999, 5)

The ability to accumulate knowledge seems to be an essential factor distinguishing man from other animal species. The ability to behave intelligently and innovatively is given not only to humans. However, while it also occurs among other primates, it is individual in their case. This means that other individuals in the group do not engage in the social learning that would allow knowledge to accumulate over time. Neither is the use of technology unique to humans, for animals are also able to use primitive tools. The difference lies in the form of acquiring knowledge and the ability to accumulate it. As Ray

Kurzweil notes, "Other animals communicate, but they don't accumulate an evolving and growing base of knowledge to pass down to the next generation" (Kurzweil 2005, 329). In the light of the epistemological system of politics, this indicates that social cognition seems to be a more adequate premise for the anthropological uniqueness of human behaviour than using tools. And the very concept of the cumulative process of cultural evolution is compatible with the theory of spontaneous order as recognising the social nature of man and indicating the existence of limitations in the conscious and planned formation of social reality.

In the context of sociological premises, the theory of spontaneous order fits into the processual approach. Due to the key role of human action in the perception of the social world, this theory creates a dynamic picture of social phenomena in which institutions do not exist beyond the meanings given to them by individuals. This means that structural interdependencies and relations between individuals result from actions taken by people. Action leads to the creation and change of social structures, which themselves create the context for future actions. So the theory adopts an event-based approach that sees society as an ongoing process in which "individual communities undertake culturally designated and structurally oriented activities towards each other and in the course of this process modify and produce groups, social systems, social structures and culture, which in turn provide a context for future activities" (Sztompka 2012, 36). At the same time, due to the attribution of purposefulness and causality to the individual resulting from the concept of human action, the theory of spontaneous order opposes both functionalism and structuralism. Because man is reflexive and his actions are nondeterministic, social processes are open, which leads to the rejection of the view of society as striving for a state of equilibrium. And the endogenous description of social changes presented by the theory fits in with the criticism of both these approaches, drawing attention to their difficulties in explaining the origin of these changes.

In the context of the scientific model, the theory of spontaneous order seeks to explain social reality in terms of human action. In this perspective, human action is the causative factor of social phenomena. The phenomena themselves are a construction based on the beliefs possessed by individuals and manifested in their actions. This means that all institutional constructs, aggregates and statistical quantities cannot be real causes of the observed phenomena, because they themselves are an effect of the interpretation of human activities. Hence, it is impossible to predict the future state of social reality. Human behaviour, and consequently future social phenomena, depend on the knowledge possessed by individual persons. Due to the anti-naturalism of the theory, the individual's beliefs are a final cause and are nondeterministic. This means that regularities and relationships described by these constructs are not scientific laws. While such relationships may appear, nondeterministic changes in the beliefs of individuals in action make them impermanent. They only provide some historical knowledge about a given time and place. This means that, regardless of how we theorise these constructs, there are no cognitively available laws governing complex social structures. At the same time, perceiving the beliefs and motivations of individuals to the ultimate cause of social phenomena turns their explanation into a reconstruction of them with the use of the composite method.

In the context of axiological premises, the theory of spontaneous order strongly emphasises the importance of liberally conceived individual freedom. This corresponds to the methodological assumption of individualism, which, although it is not a necessary or sufficient condition for the political attitude of individualism and liberalism, is correlated with it (Udehn 2001, 337–338). In particular, the freedom and autonomy of the individual assumed by liberalism correspond to the concept of human action presented by the theory under discussion. Characteristically, however, unlike other trends in contemporary liberalism, the issue of freedom is not derived here from ethical premises, but mainly from the adopted theory of society (cf. Nozick 1974; Rawls 1971). This means that this concept does not stem from assumptions about the existence of fundamental natural rights or inalienable human rights.[5]

This normative aspect of the theory of spontaneous order is mainly based on an epistemological argument, invoking the existence of immanent human cognitive limitations and, consequently, constraints on the possibility of a planned formation of social reality. In the context of the dispersed nature of knowledge, this means a limited ability to aggregate the information needed by a central governing body to achieve its goals. And so the key to coordinating the activity of individuals is for them to have the freedom of action that enables the transmission and creation of information in the course of interactions between them. This in turn makes it possible to use knowledge possessed by others, to which the individual does not have direct access. This reliance on the epistemological argument means that the key aspect here is not the right to be free in itself, as is the case with approaches based mainly on ethical considerations, but how this freedom serves all members of society. As Hayek wrote:

> It is not because we like to be able to do particular things, not because we regard any particular freedom as essential to our happiness, that we have a claim to freedom. The instinct that makes us revolt against any physical restraint, though a helpful ally, is not always a safe guide for justifying or delimiting freedom. What is important is not what freedom I personally would like to exercise but what freedom some person may need in order to do things beneficial to society. This freedom we can assure to the unknown person only by giving it to all.
>
> (Hayek 1960, 84)

This approach seems to provide an attractive argument for the liberal concept of freedom. Contrary to positions based on ethical considerations, it avoids problems with trying to infer the role of social institutions (e.g., the state) from moral norms. Moreover, it makes it possible to defend freedom so conceived also on utilitarian grounds.

As in the case of ideological premises within political science, adopting a liberal perspective is connected with the need for a legal system warranting individual freedom. Violation of this freedom affects the ability to freely interact and transmit information between people and may lead to a curtailing of the bottom-up coordination of individual activities. Embedding the idea of freedom in human cognitive limitations also makes the theory of spontaneous order strongly emphasise the negative effects of law-making by the state. State legislation is a manifestation of the planned shaping of social reality, which is constrained by human cognitive limitations.[6] This prompts some researchers to formulate normative conclusions, questioning the government's rights to legislate (see Rothbard 1973; Benson 2011; Leeson 2014). This makes the theory of spontaneous order ideologically related to libertarianism, for which it forms part of the theoretical foundation (Hamowy 2008, 488).

The normative order is so complex that our theory regards it as spontaneous, and in the long-term perspective seen as produced by biological and then cultural evolution. In this view, a norm or principle means a tendency to act in a certain way (Hayek 1973, 76). This ontological reduction of norms is based on the assumption that institutions are beliefs in the mind of the agent. But following from the anti-naturalism of the theory, knowledge about these norms is available indirectly, through the human practices and habits in which it manifests itself. Consequently, the understanding of law and normative order in the theory of spontaneous order is compatible with the approach of legal realism (Gorazda 2012). This current of the philosophy of law promotes a substantive understanding of law, treating it as an actually (empirically) existing reality (Stelmach 1995, 39). This means that it is not the formal norms created by the legislator that constitute the law, but the legal practice and its effects – the work of the courts, expectations of participants or effects of prevailing norms.

Notes

1 Cf. Godlewski 2010, 141–142.

2 This trend is undoubtedly influenced by the issue of expanding the functions of the state in various areas of social life.

3 In particular, if one accepts the understanding of politics as a process of exerting influence, this problem manifests itself in the ambiguity of the concept of influence. For every interaction between individuals, as having an impact on the decisions of others, is a form of exerting influence.

4 Klementewicz (2014, 278) uses the term "the legal state" instead of "the rule of law". It should be noted that although the concepts behind both phrases are in many ways akin to each other, they are not the same (Krygier 2014, 46–48; Sharandin and Kravchenko 2014; cf. Kriste 2014). The concept of the legal state (German *Rechtsstaat*, French *état de droit*, Spanish *estado de derecho*) indicates the strict relational connection between the law and the state and is characteristic of continental tradition and legal thought. In turn, the term "rule of law" is rooted in the Anglo-Saxon tradition of precedent law. Thus, the distinction between the notions of "the legal state" and "the

rule of law" refers to the distinction between the tradition of continental law and the tradition of common law present in legal culture. In particular, they differently perceive the relationship between the state and the law, as well as conceiving of the concept of civil society differently (Izdebski 2010, 241–243). In the Anglo-Saxon tradition, the idea of the rule of law strongly emphasizes the primacy of law over people, including those in power. This requires that it be, as far as possible, separated from the will of both individuals and communities, and especially from the legislature. In turn, the idea of the legal state focuses on the role of the law as a limitation of the will of public authority. It largely refers to legal positivism, recognising the fundamental role of the legislature in the creation of law. Thus, taking into account the context in which the term is used, and especially the concept of a democratic-liberal state, it seems more appropriate to use the term "rule of law".

5 The opposite stance, embedding the concept of freedom in morality, can be found in the work of, among others, Robert Nozick (1974, 6): "Our starting point then, though non-political, is by intention far from nonmoral. Moral philosophy sets the background for, and boundaries of, political philosophy".

6 The issue of the significance and role of the legislative as presented by the theory of spontaneous order is discussed in more detail in Chapter 7.

Bibliography

Antoszewski, Andrzej, and Ryszard Herbut. 2003. *Leksykon politologii*. Wrocław: Wydawnictwo Alta2.

Benson, Bruce L. 2011. *The Entreprise of Law: Justice without the State*. Oakland: The Independent Institute.

Beyme, Klaus von. 2006. *Die politischen Theorien der Gegenwart: Eine Einführung*. 7th ed. Opladen: Westdeutscher Verlag.

Blok, Zbigniew. 2009. *O polityczności, polityce i politologii*. Poznań: Wydawnictwo Naukowe WNPiD UAM.

Blok, Zbigniew, and Małgorzata Kołodziejczak. 2015. O statusie i znaczeniu kategorii 'polityki' i 'polityczności' w nauce o polityce. *Studia Politologiczne,* 37: 17–32.

Borowiec, Paweł. 2014. Konsekwencje twierdzenia, że 'to nie przedmiot konstytuuje teorię, ale teoria swój przedmiot' oraz czy badanie zjawisk politycznych może się obyć bez wstępnej teorii? In *Odmiany współczesnej nauki o polityce*, edited by Piotr Borowiec, Robert Kłosowicz and Paweł Ścigaj, vol. 1, 3–29. Kraków: Wydawnictwo Uniwersytetu Jagiellońskiego.

Dahl, Robert A. 2004. Complexity, Change and Contingency. In *Problems and Methods in the Study of Politics*, edited by Ian Shapiro, Rogers Smith, and Tarek Masoud, 377–381. Cambridge: Cambridge University Press.

Dahl, Robert A., and Bruce Stinebrickner. 2002. *Modern Political Analysis*. 6th ed. Englewood Cliffs, NJ: Prentice-Hall.

Feyerabend, Paul. 1993. *Against Method*. London: Verso.

Giddens, Anthony. 1987. *Social Theory and Modern Sociology*. Stanford: Stanford University Press.

Godlewski, Tomasz. 2010. Politologia a socjologia – wpływ wzajemnych relacji na tożsamość dyscypliny. *Athenaeum* 26: 139–146.

Goody, Jack. 1986. *The Logic of Writing and the Organization of Society*. Cambridge: Cambridge University Press.

Gorazda, Marcin. 2012. Normativity According to Hayek. In *The Many Faces of Normativity*, edited by Jerzy Stelmach, Bartosz Brożek, and Mateusz Hohol, 223–256. Kraków: Copernicus Center Press.

Hamowy, Ronald. 2008. *The Encyclopedia of Libertarianism*. Los Angeles, London: SAGE Publications.

Hayek, Friedrich A. 1960. *The Constitution of Liberty*. Chicago: University of Chicago Press.

Hayek, Friedrich A. 1973. *Rules and Order*, Vol. 1 of *Law, Legislation and Liberty*. London: Routledge.

Izdebski, Hubert. 2010. *Doktryny polityczno-prawne: fundament współczesnych państw*. Warszawa: LexisNexis.

Karwat, Mirosław. 2006. Przedmiot badań politologicznych w świetle zasady aspektowości. In *Teoretyczne podstawy socjologii wiedzy*, edited by Paweł Bytniewski, and Mirosław Chałubiński, vol. 1, 38–57. Lublin: Wydawnictwo UMCS.

Karwat, Mirosław. 2009. Syndromatyczny charakter przedmiotu nauki o polityce. In *Demokratyczna Polska w globalizującym się świecie. I Ogólnopolski Kongres Politologii*, edited by Konstanty Adam Wojtaszczyk, and Andżelika Mirska, 175–188). Warszawa: Wydawnictwa Akademickie i Profesjonalne.

Karwat, Mirosław. 2010. Polityczność i upolitycznienie. Metodologiczne ramy analizy. *Studia Politologiczne* 17: 63–88.

Klementewicz, Tadeusz. 1991. *Spór o model metodologiczny nauk o polityce*. Warszawa: ISP PAN.

Klementewicz Tadeusz. 2004. Teoria polityki w praktyce badawczej. In *Współczesne teorie polityki – od logiki do retoryki*, edited by Tadeusz Klementewicz, 234–260. Warszawa: Instytut Nauk Politycznych UW.

Klementewicz, Tadeusz. 2013. Politolog w labiryncie paradygmatów – pułapki eklektyzmu. In *Podejścia badawcze i metodologiczne w nauce o polityce*, edited by Barbara Krauz-Mozer, and Paweł Ścigaj, 31–43. Kraków: Księgarnia Akademicka.

Klementewicz, Tadeusz. 2014. Poza kredowym kołem polityczności. Politologia jako wieloparadygmatyczna struktura wiedzy. In *Odmiany współczesnej nauki o polityce*, edited by Piotr Borowiec, Robert Kłosowicz, and Paweł Ścigaj, vol. 1, 269–293. Kraków: Wydawnictwo Uniwersytetu Jagiellońskiego.

Krauz-Mozer, Barbara. 1992. *Metodologiczne problemy wyjaśniania w nauce o polityce*. Kraków: Uniwersytet Jagielloński.

Krauz-Mozer, Barbara. 2005. *Teorie polityki. Założenia metodologiczne*. Warszawa: Wydawnictwo Naukowe PWN.

Krauz-Mozer, Barbara. 2009. Metodologia politologii w perspektywie pluralistycznej. In *Demokratyczna Polska w globalizującym się świecie. I Ogólnopolski Kongres Politologii*, edited by Konstanty Adam Wojtaszczyk, and Andżelika Mirska, 149–163. Warszawa: Wydawnictwa Akademickie i Profesjonalne.

Krauz-Mozer, Barbara, and Paweł Ścigaj. 2013. Sklep z podróbkami? Podejścia badawcze i metodologiczne w nauce o polityce. In *Podejścia badawcze i metodologiczne w nauce o polityce*, Barbara Krauz-Mozer, and Paweł Ścigaj, 9–27. Kraków: Księgarnia Akademicka.

Kriste, Stephan. 2014. Philosophical Foundations of the Principle of the Legal State (Rechtsstaat) and the Rule of Law. In *The Legal Doctrines of the Rule of Law and the Legal State (Rechtsstaat)*, edited by James R. Silkenat, James E. Hickey Jr., and Peter D. Barenboim, 29–43. Cham: Springer.

Krygier, Martin. 2014. Rule of Law (and Rechtsstaat). In *The Legal Doctrines of the Rule of Law and the Legal State (Rechtsstaat)*, edited by James R. Silkenat, James E. Hickey Jr., and Peter D. Barenboim, 45–59. Cham: Springer.

Kurzweil, Raymond. 2005. *The Singularity Is Near: When Humans Transcend Biology*. London: Penguin Books.

Leeson, Peter T. 2014. *Anarchy Unbound: Why Self-Governance Works Better Than You Think*. New York: Cambridge University Press.

Nocoń, Jarosław. 2010. Problem granic dyscyplinarnych politologii. Athenaeum, vol. 26, 51–62.

Nozick, Robert. 1974. *Anarchy, State, and Utopia*, New York: Basic Books.

Rawls, John. 1971. *A Theory of Justice*. Cambridge, MA: The Belknap Press of Harvard University Press.

Robbins, Lionel. 1932. *An Essay on the Nature and Significance of Economic Science*. London: Macmillan.

Rosicki, Remigiusz, and Wiktor Szewczak. 2012. O przedmiocie badań politologii. Czy możliwa jest ogólna teoria polityki? *Studia Polityczne*, no. 29: 37–62.

Rothbard, Murray N. 1973. *For a New Liberty: The Libertarian Manifesto*. New York: Macmillan.

Rothbard, Murray N. 2009. *Man, Economy, and State with Power and Market*. Auburn: Ludwig von Mises Institute.

Sharandin, Yury A., and Dmitry V. Kravchenko. 2014. Rule of Law, Legal State and Other International Legal Doctrines: Linguistic Aspects of Their Convergence and Differentiation. In *The Legal Doctrines of the Rule of Law and the Legal State (Rechtsstaat)*, edited by James R. Silkenat, James E. Hickey Jr., and Peter D. Barenboim, 145–151. Cham: Springer.

Stelmach, J. 1995. *Współczesna filozofia interpretacji prawniczej*. Kraków: Wydział Prawa i Administracji Uniwersytetu Jagiellońskiego.

Sztompka, Piotr. 2012. *Socjologia. Analiza społeczeństwa*. Kraków: Wydawnictwo Znak.

Ścigaj, Paweł. 2010. Granice w politologii jako wyznacznik tożamości dyscypliny. *Athenaeum*, vol. 26, 32–50.

Tomasello, Michael. 1999. *The Cultural Origins of Human Cognition*. Cambridge, MA: Harvard University Press.

Udehn, Lars. 2001. *Methodological Individualism. Background, History and Meaning*. London: Routledge.

Wierciński, Andrzej. 1997. *Magia i religia: szkice z antropologii religii*. Kraków: Nomos.

Woleński, Jan. 1982. Dyscyplina naukowa a teoria naukowa. *Zagadnienia Naukoznawstwa* 69–70 (1–2): 3–11.

6 The theory of spontaneous order versus public choice theory

Public choice theory can be described as a study of political phenomena using the tools of the economic sciences (Tullock 2008, 723), or as an economic study of non-market decision-making processes (Mueller 2003, 1).[1] It belongs to the current of new institutional economics, relying on the neoclassical approach in economics. The main areas of research are the theory of social choice, constitutional economics, the theory of public goods, and the theory of interest groups and rent-seeking. The first of these areas concerns methods of collective decision-making and the social evaluation of selected solutions. Constitutional economics examines the diversity of power structures using an economic analysis of the constitutional order, while also seeking effective solutions for the organisation of government structures. Research in the third area focuses on the issue of using and effectively delivering public goods. The last theory deals with the motivations and operating modes of interest groups and the related rent-seeking effect. In the normative aspect, all these areas of study serve to assess the economic effectiveness of the institutional system of society and determine the ways in which it can be optimised. In particular, public choice theory suggests that transaction costs occur not only in the sphere of market activities, but also in the context of public choices. As a result, it seeks to optimise the relationship between private and public realms.

Adopting public choice theory as an important reference point for considering the use of the theory of spontaneous order in political science seems to be advisable for several reasons. In particular, there is a strong link between the two research approaches and the economic sciences, as well as some affinity of methodological assumptions. In the case of the theory of spontaneous order, this connection results from the origins of the theory and its embedding in the methodology of the Austrian School of Economics. Public choice theory, on the other hand, is mainly based on the neoclassical doctrine in economics. Despite significant differences between the two approaches, both are based on a form of methodological individualism, which makes this approach universal and applicable in various areas of social science.

The above observations suggest that the spontaneous order approach may be used in political sciences to analyse issues raised by the public choice theory, to propose up-to-date solutions and to pinpoint new problems to be scrutinised by this theory.[2]

The problem of knowledge

Public choice theory

Unlike the theory of spontaneous order, the traditional public choice approach is largely based on methodological assumptions of neoclassical economics.[3] Despite some similarities, these assumptions differ from those adopted by the Austrian School of Economics, so these approaches adopt different perspectives on the phenomena studied. But this also suggests that the criticism of neoclassical economics by the Austrian School may be transplanted to political science. This in turn makes it possible for the theory of spontaneous order to identify problems to be analysed by public choice theory and to propose new ways of solving them.

While particular approaches within public choice theory show some differences, their common denominator is the use of the neoclassical model of market behaviour to describe political phenomena (see Evans 2014, 26–29; Ikeda 2003, 68–69). This model uses the concept of rational choice, based on methodological individualism (see Buchanan 1990, 1), according to which individuals choose from available options on the basis of their preferences in order to achieve maximum utility. This approach assumes that individuals possess all the knowledge needed to determine this maximum and thus to make an optimal decision.[4] It means that this knowledge, referred to as "perfect knowledge", is objective and unquestionable (certain).[5] It follows that choice is based on a calculation, taking into account the probabilities of events, and hence the profits and costs objectively connected to them.

From the point of view of such rationality, choices made within the market are no different from non-market choices. In both cases, the agents, acting in accordance with individual preferences, seek to achieve individual benefits. So the behaviour of individuals is motivated by their own interest. Reliance on motivation as a basis for explaining political phenomena leads public choice theory to question the assumption that political actors can be altruistic. Buchanan (2003) describes such an approach as "politics without romance". In particular, this means rejecting the perception of the behaviour of politicians and bureaucracy as based on altruistic motives (Downs 1957; Niskanen 1994; Tullock 2005). Public institutions behave in a similar way to private companies, pursuing their own development and growth. Given identical motivations, what makes private and public choice different is the institutional setting within which decisions are made. This shapes and limits the possibilities of achieving individual benefits. This gives public choice theory a normative aspect, suggesting that it is not the benevolence of the actors, but the appropriate choice of constitutional norms that constitutes an essential factor in the success of political reforms. As Buchanan pointed out, "scientists, as analysts of politics, should spend more time in inquiry about the workings of different rules and less on efforts to modify the behaviour of those in the roles of political agents. Changing the rules is perhaps much easier than changing the character of the players" (Buchanan 2008, 178).

Grounding public choice theory in neoclassical economic doctrine means that political phenomena are described by the equilibrium model. In the economic context, this model describes the market as an aggregation of choices made by rational agents who pursue the principle of the maximisation of utility. The market reaches a state of equilibrium, understood as optimum economic efficiency when supply balances out with demand. This occurs when transaction costs are zero. This state is effective in the Pareto sense, meaning that it is impossible to improve someone's situation without worsening the situation of someone else. However, since the transaction costs, i.e., those related to acquiring information and executing the exchange, are not zero, the market cannot be in a state of equilibrium. In this sense, market mechanisms are not able to achieve an efficient allocation of the goods that would guarantee an equilibrium between supply and demand, which leads to market failure. This suggests the need for government intervention in the economy. If transaction costs were zero, the market would reach the optimum in the Paretian sense by itself, making such intervention unnecessary (Coase 1960).

In public choice theory, a similar situation also occurs in the political realm. By interpreting political reality in terms of rational choice and maximisation of utility, this theory indicates that, as in the economy, interactions between actors also constitute a form of exchange in politics (Buchanan and Tullock 1962, 23). Consequently, public choice theory treats politics as a form of the market, created by the rational actions of its participants. *Homo politicus* is identified with *homo economicus*. In particular, both the free market and the government serve to coordinate the actions of individuals. This turns politics into a positive-sum game. However, every exchange process entails some transaction costs, connected not only with economic activity, but also with political life and the government. So while the operations of government structures serve to overcome market inefficiencies, they may themselves be incapable of achieving the adopted goals. Echoing the concept of market failure, this constitutes a government failure, understood as its inability to operate effectively, meaning both an ineffectiveness of actions taken and absence of any action (Orbach 2013, 56).

The primacy of the problem of knowledge

The theory of spontaneous order's rejection of the neoclassical perspective leads to significant differences between this approach and public choice theory. While both describe social phenomena in terms of individuals pursuing their own goals, they present different concepts of human action. The neoclassical foundation of public choice theory assumes the availability of knowledge about the means and goals of action and so it equates human conduct with a passive reaction of maximising utility within the framework of defined and known limitations. But from the point of view of the spontaneous order theory, man acts in conditions of cognitive limitations and hence of incomplete knowledge. This means that he acts in a state of uncertainty. This assumes the existence of creativity, meaning that human conduct is not about using given means to achieve given goals, but about discovering new means and goals. In this sense, human creativity produces the difference between action and passive reaction described by public choice theory. Individuals, using experience and entrepreneurship, actively shape their future. And the subjective nature of human knowledge means that that it is knowledge rather than objective external factors that imposes limitations on human activity. This difference leads to various understandings of information. In neoclassical terms, information is objective and like other goods it provides a means of human action. In the theory of spontaneous order, information constitutes subjective and often inarticulate knowledge created, used and interpreted in the context of specific conduct.

The above comparison allows us to point out two key aspects distinguishing the theory of spontaneous order from public choice theory: the issue of imperfect knowledge[6] and the issue of creativity. It also suggests that the key aspect in the study of political and social phenomena for the spontaneous order theory is not individual motivation, but the availability of knowledge. In contrast, public choice theory, as based on the concept of rational choice, perceives political order as an intended outcome of decisions made by individual actors: "People get what they want". However, since perfect knowledge rules out unplanned order formation, any explanation of government failure must invoke the motivation of individual actors. In contrast, the theory of spontaneous order allows for the existence of unintended results of human activity. Consequently, it expands the research field through identifying mechanisms of unintentional order creation. It does not deny the possibility of selfish behaviour on the part of political actors. In this theory, based on the category of human action, man always strives for satisfaction. As this desire is shaped by subjective and irreducible individual preferences; it is the ultimate cause of action. Since these preferences are subjective, the theory of spontaneous order assumes that motivations guiding human actions may vary from person to person, so they are not limited to altruistic attitudes (cf. Ikeda 2003, 66).

Due to these discrepancies, the two theories focus on different areas of human activity and assume a different starting point for their research (Ikeda 2003, 65). In the case of public choice theory, it is the discrepancy between the declared and actual intentions of the agent. In contrast, the starting point for the spontaneous order theory is the existence of cognitive limitations accompanied by an unrestricted range of motivations to act. This produces discrepancies between the expected and actual effects of human actions.

Importantly, focusing on the incompleteness of the knowledge possessed by individuals means not only adopting a different perspective on the examined phenomena, but also recognising the approach based on motivation as wrong. Motivation to act is a derivative of an individual's view of the world. It results from how an agent perceives and understands the surrounding reality.

Motivation thus depends on cognitive processes and the knowledge created in their course. This approach, described by Anthony Evans as the "Epistemic Primacy Thesis" (2014, 24), suggests the predominance of the knowledge problem. Consequently, the issue of motivation, as well as ideas and beliefs possessed by the agents, is just a concomitant of the problem of knowledge and has no independent significance beyond it. Moreover, this shows that since public choice theory is based on the equilibrium model, it does not take into account the endogenous factor of change in the form of human creativity and entrepreneurship.

The epistemic primacy of the problem of knowledge also allows us to identify the causes of "imperfections" in social and political systems. In his classic book *The Road to Serfdom* (1944), Hayek points out that the origins of authoritarian and totalitarian systems can be found not so much in the evil intentions of the people creating them, as in the fatal faith in the ability of human reason to shape social reality, often with the best intentions. Moreover, it is precisely good intentions, more than bad ones, that can threaten the liberal concept of freedom when combined with the conceit of reason, because the former carry much more weight in society. In this sense, the idea of the primacy of the problem of knowledge leads to a conclusion which is well illustrated by the proverb "the road to hell is paved with good intentions". Even if political agents act for altruistic reasons, they do not necessarily have access to the knowledge needed to carry out their ambitious plans. So it can be seen that the implementation of central control of the economic system and the centralization and accumulation of political power does not result from a willingness to arbitrarily use violence, but is the cause of such willingness (see Boettke 1995).

Implications

Radical ignorance

Imperfection of knowledge, one of the two aspects distinguishing the spontaneous order and public choice approaches, leads to a rejection by the former of the concept of rational choice on which public choice theory is based. However, while the neoclassical model of rationality is criticised for its atomic nature and lack of empirical confirmation (see Green and Shapiro 1994), its rejection by the spontaneous order approach mainly follows from the question of the cognitive faculties of agents.

Both the theory of spontaneous order and public choice theory assume that human action is rational and founded on the existence of a logical structure of mind common to all people and providing the basic theoretical framework for interpreting this conduct. However, the similarity ends here. The theory of spontaneous order, rejecting the neoclassical assumption about the availability of knowledge of means and goals, assumes that human action is inseparable from uncertainty. Consequently, this model does not see rationality as embedded in the concept of perfect knowledge. This leads to a critique of the two main approaches describing the mechanism of human choice in public choice theory.

These mechanisms, called rational ignorance and rational irrationality, attempt to reconcile neoclassical rationality with observed mistakes in human action. The first suggests that ignorance leading to mistakes results from the fact that the costs of acquiring the knowledge needed to avoid mistakes are greater than the resulting benefits (Wittman 1989). In this sense, the lack of knowledge and the resulting mistakes follow from a rational choice – rational in the sense of being based on a calculation of profit and loss. As for rational irrationality, this is based on the distinction between epistemic and instrumental rationality. The former means constructing true beliefs and avoiding false ones, while the latter means choosing methods for achieving goals in accordance with one's beliefs. As Caplan says (2001), people have preferences in the realm of beliefs. This makes having a false belief (being epistemically irrational) rational as long as the costs involved are outweighed by the need to have that belief because of one's preferences. This means that while rational ignorance combines both forms of rationality, rational irrationality concerns situations where instrumental rationality is epistemically irrational.

The difference between the two approaches is based on the subjective beliefs of particular actors. By adopting the neoclassical approach, rational ignorance sees human action as resulting from the calculation of objective costs and benefits, regardless of the subjective beliefs of individuals. And the concept of rational irrationality, distinguishing between epistemic and instrumental rationality, leaves room for the influence of individual subjective preferences on people's choices.

Criticism of the concept of rational ignorance from the perspective of the theory of spontaneous order is based on its failure to see the cognitive limitations of man. It assumes that knowledge of the costs and benefits associated with accepting and possessing particular views is objectively given. This regards all possible beliefs. The resulting full knowledge of individuals about costs and benefits means that the agent's choices and future actions are predetermined. This makes it possible to describe reality through a model of equilibrium, in which all factors are given, thus denying the existence of any problem to be solved. The same applies to the concept of rational irrationality. Although it takes into account the role of individual subjective beliefs, it does not reject the availability of knowledge about costs and benefits, thus assuming the deterministic nature of human actions.

In both concepts, cognitive inadequacy does not result from ignorance, but from a calculation based on the knowledge required (Caplan 2007, 123). And in the case of rational irrationality, maintaining erroneous views while knowing about the costs involved suggests that man deliberately avoids epistemic rationality (see also Bennett and Friedman 2008). However, to deliberately avoid a true belief, individuals must know in advance that this belief is true. Therefore, even if a person accepts a false belief, he or she knows that it is false. However, since this knowledge itself constitutes a belief, it makes the individual's belief system self-contradictory. And this is inconsistent with rationality.

Contrary to both of the above approaches, in the spontaneous order theory individual beliefs are not the result of an algorithmic process based on data and objective factors. Rejection of the assumption of perfect knowledge suggests the actors' ignorance, understood as a lack of knowledge of what is not known (see Friedman 2005; Evans and Friedman 2011). So unlike in rational ignorance, where people know they are ignorant, the theory of spontaneous order takes a position of radical ignorance.[7] This means that it does not rationalise decisions in terms of alleged gains and losses. For action takes place not so much in conditions of calculated risk as in conditions of permanent uncertainty. So this approach allows for false beliefs. However, unlike in the case of rational irrationality, people don't know about their invalidity, which cannot be included in rational calculation. It should also be noted that creativity makes individuals able to acquire new information affecting their beliefs. Consequently, decisions made by individuals are conditioned by a given context of time and place. So unlike in the idea of rational choice, individuals do not have to make the same decisions every time they are confronted with the same choices (cf. Downs 1957, 27).

Government failure

The spontaneous order theory posits the radical ignorance and creativity of individuals, leading to the rejection of the concept of market and government failure as conceived by the theory of public choice. These concepts are based on the assumption of the availability of knowledge allowing for the calculation of profits and costs. In particular, this means knowledge of transaction costs. In contrast, in the spontaneous order theory, people act in conditions of uncertainty, not having full knowledge of transaction costs and so making wrong decisions which do not result from an objective calculation. So as Israel Kirzner (1973, 225–233) noted, mistakes, rather than transaction costs resulting from radical ignorance, are the main obstacle to the coordination of individual actions and maximisation of utility. And the subjectivism of the theory means that these costs are never objectively given. It is human entrepreneurship that makes people constantly discover new ways of acting, thus influencing the level of these costs. This means that the issue of coordinating individual actions is not a problem of maximising the known utility function within a specific range. The problem is posed by the lack of knowledge of means and goals, which, rather than given and constant, is dispersed and changeable due to human ability to create information *ex nihilo*.

The problem of limitations on access to information is present not only in economics, but also in politics. In both these spheres, man is in a state of uncertainty, constantly discovering new knowledge and the resulting possibilities of action. This makes it impossible to determine the state of equilibrium, which makes the concept of failure adopted by public choice theory irrelevant.[8] Moreover, the concept is based on a comparison of the actual situation with a perfect equilibrium. Consequently, this comparison constitutes what is known as the nirvana fallacy, that is comparing imperfect reality with an idealised model overlooking the limitations of the real world (Demsetz 1969). Such a model, often in an indirect or non-verbalised way, creates unrealistic or false assumptions resulting from an idealistic view based on the researcher's subjective beliefs.[9]

Contrary to the above approach, the theory of spontaneous order rejects the assumption of perfect knowledge. Acting in conditions of uncertainty, agents make unintended mistakes. In this sense, a failure is seen not as a predicted and calculated effect of the agents' motivation, but as a discrepancy between their intended goals and achieved results. In this sense, a government failure can be understood as a situation in which the effects of its actions differ from those expected by their proponents (see Ikeda 1997, 110). This approach avoids the nirvana fallacy, as it does not compare actual results to an ideal model, but juxtaposes them with the expectations behind the actions leading to these results.

It should be mentioned that while the phenomenon of government failure is recognised by the theory of spontaneous order, it has no equivalent in the economic sphere. Founding the concept of failure on the issue of imperfection of knowledge, and thus the discrepancy of effects with intentions, means that it concerns intentional behaviour. In contrast, the market is a decentralised process, to which intentionality cannot be attributed.

Notes

1 As noted before, such analyses are present mostly in the neoclassical approach to the economic sciences.
2 Among the issues concerning the state discussed by public choice theory many were previously addressed by the Austrians (Boettke and Leeson 2004). This applies especially to issues of bureaucracy (Mises 1944) and the emergence of interest groups (Hayek 1944; Mises 1990b).
3 Several approaches can be distinguished within public choice theory (the Virginia School, the Chicago School, the Rochester School), differing in the extent to which they use the assumptions and mathematical apparatus of neoclassical economic doctrine.
4 As can be seen, the type of rationality resulting from the concept of rational choice differs from rationality as defined by the theory of spontaneous order.
5 This does not mean that this kind of knowledge provides certainty as to the future state of things. It can be placed within the probabilistic space, presenting the probabilities of particular events. However, due to the nature of the probabilistic space it still allows people to determine the optimal decision. This means that while this kind of knowledge does not allow us to determine the future state of things with absolute certainty (with the probability of *1*), it gives certainty as to the probability of its occurrence. See Ikeda 2003, 66.
6 We call knowledge imperfect when it is not perfect knowledge.

7 The term "radical" is intended to distinguish ignorance understood as leading to unintended consequences from rational ignorance (Ikeda 2003, 67).
8 Criticism of market failure is widely discussed by, among others, Cowen 1992.
9 It is worth noting that due to the subject matter of the study the nirvana fallacy is not only methodological. Perceiving an idealised model in normative terms may also lead to political decisions based on such a model.

Bibliography

Bennett, Stephen E., and Jeffrey Friedman. 2008. The Irrelevance of Economic Theory to Understanding Economic Ignorance. *Critical Review* 20 (3): 195–258. doi: 10.1080/08913810802503418.

Boettke, Peter J. 1995. Hayek's the Road to Serfdom Revisited: Government Failure in the Argument Against Socialism. *Eastern Economic Journal* 21 (1): 7–26.

Boettke, Peter J., and Peter T. Leeson. 2004. An 'Austrian' Perspective on Public Choice. In *The Encyclopedia of Public Choice*, edited by Charles Rowley and Friedrich Schneider, vol. 2, 27–32. New York: Kluwer Academic Publishers.

Buchanan, James M. 1990. The Domain of Constitutional Economics. *Constitutional Political Economy* 1 (1): 1–18. doi: 10.1007/BF02393031.

Buchanan, James M. 2003. Public Choice: Politics without Romance. *Policy* 19 (3): 13–18.

Buchanan, James M. 2008. Same Players, Different Game: How Better Rules Make Better Politics. *Constitutional Political Economy* 19 (3): 171–179. doi: 10.1007/s10602-008-9046-4.

Buchanan, James M., and Gordon Tullock. 1962. *The Calculus of Consent: Logical Foundations of Constitutional Democracy*. Ann Arbor: University of Michigan Press.

Caplan, Bryan. 2001. Rational Irrationality and the Microfoundations of Political Failure. *Public Choice* 107 (3–4): 311–331. doi: 10.1023/A:1010311704540.

Caplan, Bryan. 2007. *The Myth of the Rational Voter: Why Democracies Choose Bad Policies*. Princeton: Princeton University Press.

Coase, R. H. 1960. The Problem of Social Cost. *Journal of Law and Economics* 3: 1–44. doi: 10.1002/sres.3850090105.

Cowen, Tyler, ed. 1992. *Public Goods and Market Failures: A Critical Examination*. New Brunswick: Transaction Publishers.

Demsetz, Harold. 1969. Information and Efficiency: Another Viewpoint. *Journal of Law and Economics* 12 (1): 1–22. doi: 10.1086/466657.

Downs, Anthony. 1957. *An Economic Theory of Democracy*. New York: Harper and Row.

Evans, Anthony J. 2014. A Subjectivist's Solution to the Limits of Public Choice: Resasseting the Austrian Foundations of Subjectivist Political Economy. *The Review of Austrian Economics* 27 (1): 23–44. doi: 10.1007/s11138-013-0227-7.

Evans, Anthony J., and Jeffrey Friedman. 2011. 'Search' vs. 'Browse': A Theory of Error Grounded in Radical (not Rational) Ignorance. *Critical Review* 23 (1–2): 73–104. doi: 10.1080/08913811.2011.574471.

Friedman, Jeffrey. 2005. Popper, Weber, and Hayek: The Epistemology and Politics of Ignorance. *Critical Review* 17 (1–2): 1–58. doi: 10.1080/08913810508443623.

Green, Donald, Ian Shapiro. 1994. *Pathologies of Rational Choice Theory: A Critique of Applications in Political Science*. New Haven: Yale University Press.

Hayek, Friedrich A. 1944. *The Road to Serfdom*. London: Routledge and Kegan Paul.

Ikeda, Sanford. 1997. *Dynamics of the Mixed Economy: Toward a Theory of Interventionism*. London: Routledge.

Ikeda, Sanford. 2003. How Compatible are Public Choice and Austrian Political Economy? *The Review of Austrian Economics* 16 (1): 63–75. doi: 10.1023/A:1022909308090.

Kirzner, Israel M. 1973. *Competition and Entrepreneurship*. Chicago: University of Chicago Press.

Mises, Ludwig von. 1944. *Bureaucracy*. New Haven: Yale University Press.

Mises, Ludwig von. 1990b. The Clash of Group Interests. In *Money, Method and Market Process*, edited by Richard M. Ebeling, 202–214. Boston: Kluwer Academic Publishers.

Mueller, Dennis C. 2003. *Public Choice III*. Cambridge: Cambridge University Press.

Niskanen, William A. 1994. *Bureaucracy and Public Economics*. Aldershot: Edward Elgar Publishing.

Orbach, Barak. 2013. What Is Government Failure. *Yale Journal on Regulation Online* 30 (44): 44–56.

Tullock, Gordon. 2005. *Bureaucracy*. Indianapolis: Liberty Fund.

Tullock, Gordon. 2008. Public Choice. In *The New Palgrave Dictionary of Economics*, edited by Steven N. Durlauf, and Lawrence E. Blume, vol. 6, 722–727. New York: Palgrave Macmillan.

Wittman, Donald. 1989. Why Democracies Produce Efficient Results. *The Journal of Political Economy* 97 (6): 1395–1424. doi: 10.1086/261660.

7 Limits of planning

Market processes and political processes

The approach adopted by the spontaneous order theory allows it to distinguish two aspects of the problem of knowledge. The first, resulting from the rejection of the concept of perfect knowledge, is the limited nature of human possibilities in the planned formation of social reality. In particular, this is expressed in the concept of government failure adopted by this theory. The second aspect is the possibility of spontaneous order creation. While the cognitive faculties of individuals are limited, they allow them to adapt their conduct to local conditions. This means that it is possible to make use of the dispersed knowledge possessed by other people. In the course of their activities and interactions with others, individuals discover and create new information that allows them to find new and more effective ways of achieving their goals. So human entrepreneurship provides a basis for understanding the coordinating tendencies emerging from a continuous process of interaction between individuals.

The above aspects of the problem of knowledge suggest that the spontaneous order theory raises the issue of human limits in the creation of planned order. In other words, to what extent does human rationality allow the planned formation of social reality?

With regard to the market sphere, the Austrian School of Economics points out that the mechanisms producing it allow individuals to exchange and create information, which in turn leads to mutual coordination. When taking action, individuals decide to use their resources in a manner consistent with their subjective beliefs. The resulting process of goods allocation provides people with information about the evaluation of particular goods by other market participants. This information would be unavailable without the free use of these goods by individuals possessing them. In the context of human entrepreneurship, information leads to discovering and seeking new ways of achieving profit through the pricing system. The resulting changes in the valuation and allocation of goods lead to the creation of new information, constantly driving forward the process of entrepreneurial coordination.

The Austrians also point out that human cognitive limitations significantly curtail the ability of a person to deliberately influence economic processes, because the knowledge needed to control these processes is dispersed and subjective. In addition, the dynamic nature of entrepreneurship means that, as possessed by individual agents, this knowledge is constantly changing. Consequently, it is not possible to aggregate it in order to implement an intentional plan. And due to the subjectivity of knowledge, interference in the free market mechanism reduces the scope for the effective allocation of goods. This is because it limits the ability of individuals to freely decide on the allocation of their resources, distorting the process of creating and transmitting information. This means that it prevents the valuation of resources in accordance with the individuals' preferences and, consequently, economic calculation. This in turn leads to errors and limits the ability to coordinate actions. In particular, it also makes economic calculation impossible in the socialist system (Mises 1990a), because in a system based on public ownership of the means of production it becomes impossible to evaluate them. So it deprives the central decision-making body of information on the basis of which it could assess the economic profitability of its decisions.

Observations of the market economy allow us to see how interactions between individuals lead to the spontaneous formation of the price system and the institution of money, enabling the use of the dispersed knowledge of market participants and, consequently, economic coordination. As Kirzner noted, "It is upon this spontaneous order that the unprecedented prosperity of market economies rests" (Kirzner 1987, 45). This raises the question of the extent to which both planned and unintended coordination are also possible outside the economic area.

Spontaneity can no doubt be attributed to many non-market phenomena. This applies in particular to unplanned institutions created in the process of social evolution, such as language or law. However, attention should be paid to the role of the institution of private property in the process of the spontaneous coordination of market activities, enabling the use of dispersed knowledge possessed by individuals. It is this institution that makes it possible for individuals to own goods, and thus to allocate them according to their preferences. And this determines the formation of a price system, and consequently the possibility of calculating profits and losses. The importance of private property for the coordination of economic processes is further emphasised by the negative impact of attempts to interfere with and control market processes, which are nothing other than limitations on or denials of property rights.

The above considerations allow us to observe that the issue of planned and spontaneous order regards not so much the non-economic sphere, but the sphere outside the institutional framework of property law. While market processes (based on the institution of private property) provide the information needed to coordinate individual actions, which would remain unavailable outside these processes (i.e., in a central-planning situation), it is not always possible or cost-effective to place a given good within the institutional framework of private property. In such a situation, the good becomes a public good.[1] This means that the question concerns the extent to which public goods can be created spontaneously within polycentric systems. One special public good, namely the law, seems particularly relevant here. It is the law that forms the basis for the functioning of the social order. In particular, it determines the existence of the institution of private property itself.

The institutional conditioning of economic processes suggests that the ability of a polycentric system, i.e., the free market, to coordinate individual actions is not easily extrapolated beyond the area of private-property based interactions (Kirzner 2000; Buchanan 2011). The complexity of the social world means that many of its phenomena can be seen as unintended. However, only the institutional framework of private property creates the mechanism of adjusting prices, allowing for the transformation of individual preferences into an objectively measurable variable, and thus making economic calculation possible. Outside this framework, there is no analogous mechanism that would link a variable from social reality to a variable from a social convention based on it (Boettke 2014, 241–242). So it is impossible to make a choice and act on the basis of this calculation of profits and losses. Kirzner illustrates this situation with the example of the institution of language:

> So that it is indeed useful that the language my children learn at home overlaps with the language learned by other children in their homes. This permits social intercourse and facilitates education. But there is hardly – in the insight that such institutions emerge spontaneously – any implication that the emerging institutions are the best conceivable such institutions. There is no guarantee that the English language my children learn at their mother's knee will be a "better" language for purposes of social intercourse than, say, French-or Esperanto. The demonstration that widely accepted social conventions can emerge without central authoritarian imposition does not necessarily point to any optimality in the resulting conventions.
> (1987, 48)

A similar approach is presented by Buchanan (1977). Although he points to the decentralised process in a market economy as the source of its effectiveness in the Paretian sense, he also rejects the ability of these processes to guarantee the effectiveness of all forms of social evolution.

The above criticism points to a weakness of a simple extrapolation from the coordinating capacity of market mechanisms to all social processes. However, it cannot be accepted as an argument for the inability of polycentric systems to form order outside the economic sphere. The absence of reliance on the institution of private property does not mean that the coordination of individual activity and order formation is only possible through planned and deliberate actions. Behind the creation of spontaneous order is not so much a specific institutional framework as the dispersed and subjective nature of knowledge and human entrepreneurship; it is the discovery of new information by individuals and the ability to use it that determines the possibility of coordinating human actions. This means that the importance of institutions is determined by the extent to which they enable the use of dispersed knowledge and thus coordination. Moreover, the limits of market-based phenomena are not fixed and known in advance, for they depend on the costs of defining and enforcing property rights. These in turn depend on the knowledge possessed by individuals. This means that they are not data, but are discovered in the course of ongoing activities and interactions, shaping the area of market coordination possibilities.

In the light of the issue of entrepreneurship and the availability of knowledge, the creation of spontaneous order does not have to be preceded by an institutional framework guaranteeing the enforcement of the rules governing it. So such an order can function outside government structures (see Scott 2009). This situation is illustrated, for example, by customary law, a system of enforceable rules produced by customs as a result of their repeated application to disputes and problems arising in the course of human interaction (Notten 2005, 15). The law understood in this way does not come into being in a planned manner, but as an effect of recognition of judicial solutions by community members. This means that the law is not legislated, but discovered (see Leoni 1961). Unlike statute law, this system is not controlled by any institution, but is decentralised.

An example of such a law is the commercial law (*lex mercatoria*) created in Europe with the revival of trade in the 11th and 12th centuries, associated with the agrarian revolution and increased agricultural productivity (Berman 1983, 333–334). Created by merchants themselves, it was shaped by the evolution of commercial practice (Trakman 1983; Benson 1989). It was based on customs and practices that turned out to be widely used in the course of the development of trade contacts, serving to facilitate trade and make it more efficient. The resulting uniformity and standardisation of the law reduced uncertainties in trading, especially by protecting foreign traders in disputes with local entrepreneurs and against abuses of local legislation (Berman 1983, 342).[2] The development of commercial law enabled trade to flourish in the late Middle Ages. This and other examples (Stringham 1999, 2015; Anderson and Hill 2004; Benson 2011) show the potential for a spontaneous creation of legal order. Importantly, it is not limited to conditions of a high level of trust and homogeneity of a group, but also occurs between strongly heterogeneous groups (Leeson 2007b, 2009).

The above approach differs from that presented by the theory of public choice. The latter, using the neoclassical concept of rationality, assumes that the mechanism of making choices used by individuals is independent of the sphere of activity. This is because both within and outside the market area, the behaviour of agents is motivated by their own interest. Consequently, the sphere of politics can be treated as a form of the market and interactions can be regarded as processes of exchanges of goods. However, while interactions in the free market concern choices made by individuals within the framework of the rules and laws creating the market, in the political sphere they concern the choice of rules themselves. The rational conduct of individuals in accordance with their own interests makes the effectiveness of both the market and politics dependent on institutional conditions. These conditions constitute the limitations within which agents act, influencing the possibility of them achieving their goals. In the political sphere, this suggests the importance of competition within government structures as enabling people to perceive and realise profits, and thus to more efficiently allocate resources in the political market. This is illustrated by the characteristics of the federal system indicated by Buchanan (2008, 174). The decentralised nature of this system and the related competition between the partially interdependent areas of jurisdiction of various institutions favours a better response of these institutions and the political market to the demand (defined by political preferences) generated by citizens. It leads to the reduction of the opportunistic behaviour of political actors, while improving the effectiveness of public decisions and choices made within government structures.

A similar normative attitude is contained in the concept of polycentrism formulated by the so-called Bloomington School (Ostrom 2010, 2005; Ostrom 2008). The polycentric system is characterised by the existence of many autonomous institutions with overlapping areas of operation, functioning within a common system of rules.[3] This approach, partly inspired by Buchanan's constitutional economics, suggests that the overlapping of powers leads to competition between particular elements of the system (Boettke et al. 2015, 321). This creates the possibility of choosing from among overlapping legal systems (O'Hara and Ribstein 2009). This resembles market processes because if the mechanisms of the polycentric system force institutions to adapt to the choices made by individuals, then making these choices is analogous to the signalling of profit and loss in the market. It provides feedback on the evaluation of services provided by public institutions.[4] So it mobilises particular government structures to act in a way that is more effective and better satisfies the needs of citizens.

Both the approach of public choice theory and the Bloomington School suggest the possibility of using the market mechanism of competition to understand decision-making mechanisms in the public sphere. This allows us to perceive phenomena in this area as a dynamic and unplanned interaction process. However, this perspective differs from the approach of the spontaneous order theory, which is based on the concept of entrepreneurship and as such does not assume the need for an institutional framework that would enforce rules and safeguard this process. In contrast, public choice theory embeds it within state structures.

More importantly, however, any attempt to present political processes through the mechanism of competition is problematic and inadequate under the terms of the theory of spontaneous order, because the possibility of making choices, on which the competition mechanism is based, depends on institutional conditions influencing the ability of individuals to act and on the availability of knowledge. So the choice within the market, created by the private property system, is different from public choice. In particular, as the Austrian School of Economics points out, economic calculation is only possible through choices based on the institution of private property, allowing the use of the knowledge necessary for this calculation.[5]

The theory of spontaneous order suggests that the possibility of coordinating the activities of individuals does not result from the mere existence of a number of decision-making institutions, but is based on the possibility of creating and transmitting the knowledge needed for this purpose. In the private property system, competition takes place between particular individuals, i.e., between entrepreneurial and creative agents acting in accordance with their personal preferences. This is a qualitative difference from the approach of public choice theory, for which the concept of competition also applies to relations between institutions, forming complex and often strongly centralised decision-making bodies. This means that such an approach focuses on the multiplicity of elements of the interaction system rather than on the nature of these elements. Consequently, it ignores not only the fact of humans beings as sources and carriers of information, but also the question of the subject matter of decisions and the impact of these factors on the possibility of creating order. But human cognitive barriers strongly limit people's ability to deliberately influence complex social structures.

This may be illustrated by an attempt to juxtapose the market process of creating a price system with a centrally controlled economy, where setting prices for individual goods takes place in a competition between institutions responsible for various branches of the economy, which strive to achieve their own goals. In both cases, we can speak of a system of competition in which there are competing decision-makers, but these systems produce fundamentally different results. The knowledge necessary to control the economy is dispersed among individuals, so institutions responsible for central planning cannot possess it. We should also note that the subject matter of decisions is different in these two situations. While in both cases the outcome of decisions is a price system, only in the latter instance are decisions specifically aimed at determining prices.

This creates the difference between market processes and political processes: "The political process generates incentives and learning that are entirely different than what is exhibited in the competitive market process" (Boettke et al. 2007, 138). So competition between centralised decision-making bodies, limiting the ability of individuals to create and transmit information, does not imply the ability of such a system to create order. The level at which politics is similar to market is the realm of human interactions. Politics, as the economy, is a dynamic process based on entrepreneurship. However, the framework differs making politics a "peculiar form of business practice" (Wagner 2016). In this sense rather than seen as a market process it should be treated as entangled with economy.[6] As Peter Boettke et al. (2011) point out, local governments are still governments, with all the resulting constraints on their ability to implement their plans and intentions. This undermines the "market" nature of such a polycentric system and thus the claim that competition between its elements is conducive to efficiency. In particular, the institutional setting may take the form of the prisoner's dilemma, promoting actions that harm other participants.

Democracy

The above observations apply to democratic systems. Since these systems are based on granting the right to actively participate in political decision-making to as wide a group of people as possible, this favours the perception of democratic mechanisms of choice as an analogy to market mechanisms. In both, the effect depends on the preferences of all participants. As Evans (2014, 39–40) points out, the difference between these mechanisms lies in the fact that economic activity is focused on money and the price system, while politics is focused on voting. The importance of individual preferences in market processes depends on the sums people are prepared to pay for a given good. In contrast, when people vote the weight of the participants' preferences is equal, as each has one vote. Consequently, political entrepreneurs are mindful of the number of voters who demand a given political action, while in the market the focus is on the demand expressed in money (see Boettke et al. 2007). This means that the state of the market reflects the average preferences expressed in market demand. As concerns political decisions, the equal weight of votes means that in a system of majority rule the solution chosen will correspond to the one most preferred by the median voter (Holcombe 2006, 155).[7] While the median voter theorem is qualified by a number of factors, such as transaction costs or the uniformity of the election field, the distinction between average and median preferences may be a starting point for challenging the neoclassical approach in public choice theory (Wittman 1995).

Although the claim about the median voter explains a number of political phenomena, from the point of view of the spontaneous order theory it is wrong to reduce the difference between the market and the democratic system to the mechanism of choice. As suggested earlier, identifying market processes with political processes only concerns the multiplicity of decision-makers, while ignoring the issue of availability of knowledge and the subject matter of decisions. In the democratic system we are dealing with one common subject of decisions, but since it concerns complex social structures, this process is strongly curtailed by radical human ignorance. In contrast, the market involves many decision-making processes concerning various issues, often relevant only for a given decision-maker. In particular, it can be seen from this that an attempt to set prices through a system of democratic voting would be a form of central planning, very remote from market processes. While everyone would have the right to express their preferences, such a solution, by limiting the ability of individuals to make independent decisions and operate freely within the market, would prevent the use of the dispersed knowledge of individual participants to coordinate their actions.

The focus on human cognitive limitations (radical ignorance) as the grounds for distinguishing market mechanisms from democratic systems also leads to emphasising the role of beliefs and ideas through which people interpret reality. The complexity of many social phenomena means that the effects of human decisions may not be felt directly. These effects are enmeshed in a network of relations creating a given phenomenon. In the context of radical ignorance, this limits the human ability to identify causal relationships and to adequately interpret social phenomena. This also concerns political decisions which significantly affect the complex social reality. Due to the resulting state of uncertainty, decisions are largely based on social heuristics and the extrapolation of moral beliefs (Cosmides and Tooby 1994; Dennett 1996, 494–510). In the absence of knowledge, social heuristics and moral beliefs form the basis for interpreting observed processes.

This also has an impact on the issue of the revision of human beliefs. According to the model presented in Part II, the effects of actions undertaken by individuals lead to changes in their beliefs.[8] However, in the case of complex social phenomena, radical ignorance limits access to the knowledge needed in this process. This increases the importance of the beliefs held in the interpretation of these phenomena and, consequently, in the revision of these beliefs. Consequently, the assessment and revision of political beliefs will to a lesser extent result from identifying private benefits connected with possessing and changing these beliefs than in the case of beliefs whose effects directly affect the personal benefits of the individual holding them (cf. Tarko 2015, 22). For example, assuming that voting is ineffective as a means to realise their interests, people will not be guided by these interests, but by their beliefs generated in the socialisation process (Caplan 2007). Individuals' views and beliefs form the basis for interpreting the surrounding reality. The inability to directly experience the effects of a choice made means that a person has no access to information that would allow verification and possibly a change of held beliefs. Furthermore, because of the complexity of economic, social and political structures, perceiving these effects does not necessarily imply a causal link with the choice made earlier. As concerns competing jurisdiction systems, radical ignorance may lead to an inability to adequately assign responsibility to particular institutions and so undermine the mechanism of competition and accountability (Boettke et al. 2015, 329).

Criticism of the concept of the veil of ignorance

Another area of application of the theory of spontaneous order in the context of cognitive limitations is the concept of the veil of ignorance. This concept is used in political philosophy as well as in constitutional economics. It defines a mechanism that limits or rules out making decisions based on the decision-maker's own interest by introducing conditions of uncertainty as to the distribution of benefits and losses resulting from the choices made. This thought experiment consists in assuming a certain type of ignorance of the decision maker, preventing him or her from assigning benefits and losses to particular individuals. Two approaches can be distinguished (cf. Mueller 2001). The first is based on the assumption that the decision-maker is ignorant of his own identity and the attributes that characterise him. In the second case, although the decision-maker may identify with a particular person, he does not know how the benefits and losses will be distributed as a result of his decision.

John Rawls is the author who significantly contributed to the popularisation of the idea of the veil of ignorance in contemporary philosophy (1971, 2001). He uses this concept to define the original situation in which "no one knows his place in society, his class position or social status, nor does anyone know his fortune in the distribution of natural assets and abilities, his intelligence, strength, and the like" (Rawls 1971, 12). According to Rawls, this situation, which prevents anyone from promoting his or her own position, makes it possible to define the principles of justice for the basic social structure determining the possible types of social cooperation and power relations. The concept of the veil of ignorance is also applied in the area of political economics, where it serves to explain the choice of constitutional rules (Buchanan and Tullock 1962; Vanberg and Buchanan 1989; Imbeau and Jacob 2015).[9] The idea, standing behind this approach, of neutralising the influence of personal motivation and the interests of the actors on the decision-making process makes it an essential mechanism for counteracting abuses of power, alongside the system of checks and balances (Vermeule 2001). The idea of the veil of ignorance also creates space for reflection on a wide range of issues related to decisions and choices made in a state of uncertainty. Specifically, these issues are: the legitimacy of the choice of norms and the accepted concept of justice (Nozick 1974; Hayek 1976; Daniels 1989), the possibility of a lack of unanimity and its correspondence to the degree of uncertainty (Müller 1998; Sen 2009; Muldoon et al. 2014), the degree of generality of the formulated rules (Vanberg and Buchanan 1989; Buchanan and Congleton 1998; Vermeule 2001) or the willingness to take risks (Witt and Schubert 2008; Schildberg-Hörisch 2010; Frignani and Ponti 2012).

Adopting the perspective of the theory spontaneous order allows us to point out the problematic nature not of the mechanism of decision-making from behind the veil of ignorance, but of the very fact of making a choice. The veil of ignorance, understood as a mechanism for the deliberate shaping of social reality in the interests of all rather than in the self-interest of individual participants, focuses on the decision-makers' motivation. This means that while the concept of the veil of ignorance addresses the issue of the decision-makers' knowledge, it is the knowledge making up the self-interest of individuals. This means that ignorance only concerns the possibility to assign the distribution of benefits and losses associated with the decision to specific individuals. Consequently, this approach ignores the issue of the radical ignorance of the actors and the resulting state of uncertainty also in other areas of human conduct. In contrast, the epistemic primacy of the problem of knowledge in the spontaneous order theory suggests that the possibility of forming an order according to predetermined criteria is related to the availability of knowledge rather than to motivation. Due to their cognitive limitations, people make decisions in conditions of uncertainty as to the results. This produces discrepancies between the expected and actual effects of choices.

Another aspect defining the approach of the spontaneous order theory to the veil of ignorance concept is the highlighting the role of human entrepreneurship, that is the creative ability to discover new opportunities for profit. Accordingly, the means and goals on the basis of which people make decisions are discovered in the course of their actions and acquisition of new knowledge. This argues against the concept of fairness proposed by Rawls. By focusing on the issue of fairness, it assumes the availability of knowledge about goals and resources. In contrast, radical ignorance and human creativity mean that this knowledge is not given in advance, but must be discovered. And this undermines the legitimacy of the rules formulated by that theory as based on unrealistic assumptions rejecting the concept of human action and human reflexivity.[10]

This suggests an important role for the issue of human entrepreneurship and creativity in normative analyses. According to Huerta de Soto (2010b, 20), this demonstrates the need to change the perspective on moral norms. As he writes, "clearly the fundamental ethical question ceases to be how to fairly distribute 'what exists', and becomes how, in view of human nature, to best foster entrepreneurial coordination and creativity" (Huerta de Soto 2010b, 20). This is well illustrated by the so-called Lockean proviso, which sets out the principles of fair acquisition of property. Starting from the position of human self-possession, Locke indicates that an individual has the right to things that are the result of his work. But he constrains this right with a proviso stating that a person can acquire only such an amount of goods as guarantees that "there is enough, and as good, left in common for others" (Locke 1764, 217). Human creativity makes this proviso pointless, because it is based on the assumption that the amount of resources, as well as the needs and number of people, are known and constant. In contrast, creativity indicates that regardless of the problem of determining what amount of resources is "enough", no result of human action (work) existed until it was discovered or created in the course of entrepreneurial activity, so possessing it does not necessarily cause harm to others (Kirzner 1992, 223).

This indicates that the concept of the veil of ignorance faces the problem of the radical ignorance of decision-makers, and thus the likely discrepancies between the planned and actual effects of decisions. In this sense, the theory of spontaneous order signals the risk of making the mistake of the pretence of knowledge (see Hayek 1989), manifesting itself in the belief that it is not knowledge, but a conflict of interest that constrains the human capacity for the deliberate formation of social reality. In contrast, making decisions involving complex social structures is strongly limited by radical human ignorance, regardless of the motivations guiding the actions of individual agents.[11] And the possibility of order emerging spontaneously means that rules and norms need not result from a plan directed towards this goal. In particular, as Hayek pointed out, "we must completely discard the conception that man was able to develop culture because he was endowed with reason" (Hayek 1979, 156).

Instead, the spontaneous emergence of institutions entails the human ability to learn through imitation and pass this knowledge on using these institutions.

Notes

1 A public good a good that is non-exclusive and non-rivalrous in consumption. Non-exclusive means that access to a given good by third parties cannot be prevented. Non-rivalrous means that the consumption of a given good by one person does not reduce its availability to others.
2 The disputes were resolved by private courts specialising in trade-related issues (Benson 2011, 33–34).
3 An example of a polycentric system is criminal law enforcement, constituted by the work of many institutions, including the police, local and central authorities, the justice system and private companies dealing with security issues.
4 It is worth noting that this approach, like the spontaneous order theory, recognises the importance of epistemic issues. This is because it sees institutions not so much as a factor imposing restrictions on the activities of individuals, but above all as a means of mutual coordination, enabling individuals to use knowledge that they themselves do not have. See Boettke et al. 2015.
5 Exchange within the institution of private property is described as voluntary, reflecting the understanding of economic freedom by classical liberalism.
6 Recognising spheres of politics and economy as intertwined and mutually dependent Richard Wagner (2014) coined the spase of "entangled political economy".
7 A median voter is a voter whose preferred option coincides with the median preference of all voters when choosing from a uniform spectrum of available options under a majority rule voting system. See Jakubowski 2005, 53–55.
8 See Section "Between atomism and holism" of Chapter 3.
9 A discussion of the approaches and references to the concept of the veil of ignorance formulated within constitutional economics can be found in Voigt 2015.
10 See Chapter 3.
11 This relates to the issue of limitations imposed on decision-making in the context of democratic systems, discussed in Section "Democracy".

Bibliography

Anderson, Terry L., and Peter J. Hill. 2004. *The Not So Wild, Wild West*. Stanford: Stanford University Press.

Benson, Bruce L. 1989. The Spontaneous Evolution of Commercial Law. *Southern Economics Journal* 55 (3): 644–661. doi: 10.2307/1059579.

Benson, Bruce L. 2011. *The Entreprise of Law: Justice without the State*. Oakland: The Independent Institute.

Berman, Harold J. 1983. *Law and Revolution: The Formation of Western Legal Tradition*. Cambridge: Harvard University Press.

Boettke, Peter J. 2014. Entrepreneurship, and the Entrepreneurial Market Process: Israel M. Kirzner and the Two Levels of Analysis in Spontaneous Order Studies. *The Review of Austrian Economics* 27 (3): 233–247. doi: 10.1007/s11138-014-0252-1.

Boettke, Peter J., Christopher J. Coyne, and Peter T. Leeson. 2007. Saving Government Failure Theory from Itself: Recasting Political Economy from an Austrian Perspective. *Constitutional Political Economy* 18 (2): 127–143. doi: 10.1007/s10602-007-9017-1.

Boettke, Peter J., Christopher J. Coyne, and Peter T. Leeson. 2011. Quasimarket Failure. *Public Choice* 149 (1/2): 209–224. doi: 10.1007/s11127-011-9833-8.

Boettke, Peter J., James S. Lemke, and Liya Palagashvili. 2015. Polycentricity, Self-governance, and the Art and Science of Association. *The Review of Austrian Economics* 28 (3): 311–335. doi: 10.1007/s11138-014-0273-9.

Buchanan, James M. 1977. *Freedom in Constitutional Contract*. College Station: Texas A&M University Press.

Buchanan, James M. 2008. Same Players, Different Game: How Better Rules Make Better Politics. *Constitutional Political Economy* 19 (3): 171–179. doi: 10.1007/s10602-008-9046-4.

Buchanan, James M. 2011. The Limits if Market Efficiency. *Rationality, Markets, and Morals* 2 (38): 1–7.

Buchanan, James M., and Roger D. Congleton. 1998. *Politics by Principle, Not Interest: Towards Nondiscriminatory Democracy*. Cambridge: Cambridge University Press.

Buchanan, James M., and Gordon Tullock. 1962. *The Calculus of Consent: Logical Foundations of Constitutional Democracy*. Ann Arbor: University of Michigan Press.

Caplan, Bryan. 2007. *The Myth of the Rational Voter: Why Democracies Choose Bad Policies*. Princeton: Princeton University Press.

Cosmides, Leda, and John Tooby. 1994. Better than Rational: Evolutionary Psychology and the Invisible Hand. *American Economics Review* 84 (2): 327–332.

Daniels, Norman, ed. 1989. *Reading Rawls: Critical Studies on Rawls' "A Theory of Justice"*. Stanford: Stanford University Press.

Dennett, Daniel. 1996. *Darwin's Dangerous Idea*. New York: Simon and Schuster.

Evans, Anthony J. 2014. A Subjectivist's Solution to the Limits of Public Choice: Resasseting the Austrian Foundations of Subjectivist Political Economy. *The Review of Austrian Economics* 27 (1): 23–44. doi: 10.1007/s11138-013-0227-7.

Frignani, Nicola, and Giovanni Ponti. 2012. Risk vs. Social Preferences under the Veil of Ignorance. *Economics Letters* 116 (2): 143–146. doi: 10.1016/j.econlet.2012.02.002.

Hayek, Friedrich A. 1976. *The Mirage of Social Justice*. Vol. 2 of *Law, Legislation and Liberty*. London: Routledge.

Hayek, Friedrich A. 1979. *The Political Order of a Free People*. Vol. 3 of *Law, Legislation and Liberty*. London: Routledge.

Hayek, Friedrich A. 1989. The Pretence of Knowledge. The American Economic Review 79 (6): 3–7.

Holcombe, Randall. 2006. *Public Sector Economics*. Upper Saddle River: Pearson Prentice Hall.

Huerta de Soto, Jesus. 2010b. *The Theory of Dynamic Efficiency*. London: Routledge.

Imbeau, Louis M., and Steve Jacob. 2015. *Behind a Veil of Ignorance? Power and Uncertainty in Constitutional Design*. Cham: Springer.

Jakubowski, Michał. 2005. Teoria wyboru społecznego. In *Teoria wyboru publicznego. Wstęp do ekonomicznej analizy polityki i funkcjonowania sfery publicznej*, edited by Jerzy Wilkin, 46–68. Warszawa: Wydawnictwo Naukowe Scholar.

Kirzner, Israel M. 1987. Spontaneous Order and the Case for Free Market Society. In *Ideas on Liberty: Essays in Honor of Paul Poirot*, edited by Robert G. Anderson, 45–50. Irvington-on-Hudson: Foundation for Economic Education.

Kirzner, Israel M. 1992. *The Meaning of Market Process: Essays in the Development of Modern Austrian Economics*. London: Routledge.

Kirzner, Israel M. 2000. The Limits of the Market: Real and Imagined. In *The Driving Force of the Market*, Israel M. Kirzner, 77–87. New York: Routledge.

Leeson, Peter T. 2007b. Trading with Bandits. *Journal of Law and Economics* 50 (2): 303–321. doi: 10.1086/511320.

Leeson, Peter T. 2009. The Laws of Lawlessness. *Journal of Legal Studies* 38 (2): 471–503. doi: 10.1086/592003.

Leoni, Bruno. 1961. *Freedom and the Law*. Princeton: D.Van Nostrand.

Locke, John. 1764. *The Two Treatises of Civil Government*. Edited by Thomas Hollis. London: A. Millar et al.

Mises, Ludwig von. 1990a. *Economic Calculation in the Socialist Commonwealth*. Auburn: Ludwig von Mises Institute.

Mueller, Dennis. C. 2001. The Importance of Uncertainty in a Two-Stage Theory of Constitutions. *Public Choice* 108 (3/4): 223–258. doi: 10.1023/A:1017500106015.

Muldoon, Ray, and Chiara Lisciandra, Mark Colyvan, Carlo Martini, Giacomo Sillari, and Jan Sprenger. 2014. Disagreement Behind the Veil of Ignorance. *Philosophical Studies* 170 (3): 377–394. doi: 10.1007/s11098-013-0225-4.

Müller, Christian. 1998. The Veil of Uncertainty Unveiled. *Constitutional Political Economy* 9 (1): 5–17. doi: 10.1023/A:1009087329422.

Notten, Michael van. 2005. *The Law of the Somalis: A Stable Foundation for Economic Development in the Horn of Africa*. Edited by Spencer Heath MacCallum. Trenton: The Red Sea Press.

Nozick, Robert. 1974. Anarchy, State, and Utopia, New York: Basic Books.

O'Hara, Erin A., and Larry E. Ribstein. 2009. *The Law Market*. New York: Oxford University Press.

Ostrom, Elinor. 2005. *Understanding Institutional Diversity*. Princeton: Princeton University Press.

Ostrom, Elinor. 2010. Beyond Markets and States: Polycentric Governance of Complex Economic Systems. *American Economics Review* 100 (3): 641–672. doi: 10.1080/19186444.2010.11658229.

Ostrom, Vincent. 2008. *The Political Theory of a Compound Republic: Designing the American Experiment*. Lanham: Lexington Books.

Rawls, John. 1971. *A Theory of Justice*. Cambridge, MA: The Belknap Press of Harvard University Press.

Rawls, John. 2001. *Justice as Fairness: A Restatement*. Cambridge: Belknap Press.

Schildberg-Hörisch, Hannah. 2010. Is the Veil of Ignorance Only a Concept about Risk? An Experiment. *Journal of Public Economy* 94 (11–12): 1062–1066. doi: 10.1016/j.jpubeco.2010.06.021.

Scott, James C. 2009. *The Art of Not Being Governed: An Anarchist History of Upland Southeast Asia*. New Haven: Yale University Press.

Sen, Amartya. 2009. *The Idea of Justice*. Cambridge: The Belknap Press.

Stringham, Edward P. 1999. Market Chosen Law. *Journal of Libertarian Studies* 14 (1): 53–77.

Stringham, Edward. P. 2015. *Private Governance: Creating Order in Economic and Social Life*. Oxford: Oxford University Press.

Tarko, Vlad. 2015. The Role of Ideas in Political Economy. *Review of Austrian Economics* 28 (1): 17–39. doi: 10.1007/s11138-013-0246-4.

Trakman, Leon E. 1983. *The Law Merchant: The Evolution of Commercial Law*. Littleton: Fred B. Rothman.

Vanberg, Victor, and Buchanan, James. M. 1989. Interests and Theories in Constitutional Choice. *Journal of Theoretical Politics* 1 (1): 49–62. doi: 10.1177/0951692889001001004.

Vermeule, Adrian. 2001. Veil of Ignorance Rules in Constitutional Law. *Yale Law Journal*, 111 (2): 399–433. doi: 10.2307/797593.

Voigt, Stefan. 2015. Veilonomics: On the Use and Utility of Veils in Constitutional Political Economy. In *Behind a Veil of Ignorance? Power and Uncertainty in Constitutional Design*, edited by Louis M. Imbeau, and Steve Jacob, 9–33. Cham: Springer.

Wagner, Richard E. 2014. Entangled Political Economy: A Keynote Address. In *Entangled Political Economy*. Vol. 18 of *Advances in Austrian Economics*, edited by Steven Horwitz, and Roger Koppl, 15–36. Bingley, UK: Emerald Group Publishing.

Wagner, Richard E. 2016. *Politics as a Peculiar Business: Insights from a Theory of Entangled Political Economy*. Cheltenham, UK: Edward Elgar Publishing.

Witt, Ulrich, and Christian Schubert. 2008. Constitutional Interests in the Face of Innovations: How Much Do We Need to Know about Risk Preferences? *Constitutional Political Economy* 19 (3): 203–225. doi: 10.1007/s10602-008-9044-6.

Wittman, Donald. 1995. *The Myth of Democratic Failure*. Chicago: University of Chicago Press.

8 Fragile states

State dysfunction as a coordination problem

The emphasis on the dynamic nature of social processes in the theory of spontaneous order makes it particularly useful in analysing institutional changes in countries without a stable political and legal system. Consequently, this theory may be applied to the study of so-called fragile or dysfunctional states, broadening the current view on this phenomenon in political science.[1]

The concept of a fragile state is a collective term referring to state structures whose form differs from the recognised models of statehood (Ścigaj 2013, 64). This difference is expressed in the inability of state structures to perform the functions assigned to them. The concept describing this phenomenon appeared in the 1990s (Jackson 1990; Helman and Ratner 1992; Zartman 1995) in the context of weakening of central state institutions and related armed conflicts observed after the end of the Cold War in various regions of the world (Herbst 1997). The combination of a fragile state with destabilisation and conflict leads to the perception of its failure as resulting from the absence of effective state structures. Consequently, the concept defines a situation in which a given state is unable to perform its functions.[2] This produces a tendency, widely present in the analyses of the dysfunctionality of states, to perceive this issue in terms of the reconstruction of these structures (Kaplan 2008; Boege et al. 2009a, 13; Pospisil and Kühn 2016). This is further emphasised by pointing out the dysfunctionality of countries as an important factor threatening international security (Rotberg 2002).

Under the theory of spontaneous order, the emergence and functioning of the social order is the result of the mutual coordination of individual people's actions. This coordination issues from the emergence and transmission of knowledge in interactions between agents. In the context of fragile states, such a perspective leads to an emphasis on the importance of the problem of knowledge. These countries demonstrate a certain lack of coordination in social relations. This approach means that the problem of coordination contained in the concept of dysfunctionality can be expressed in the question of how individual institutional solutions influence the emergence and flow of information, and thus the ability of agents to cooperate.

The assumption of human cognitive limitations in the theory of spontaneous order makes it focus on the role of local knowledge, that is knowledge resulting from the practices and experiences of individuals belonging to a given group or community. It consists of abilities, beliefs, habits and practices. This knowledge is closely connected with local social structures, on the basis of which it is created and passed on.[3] This highlights the role of endogenous institutions, resulting from the activities of individuals interacting with each other. As created in the process of experience formation, these institutions are rooted in local knowledge emerging from this process. They reflect the informal and often inarticulate nature of this knowledge. This also means that they are subject to change due to the altering preferences and beliefs of the individuals forming them. Local knowledge is not static and is subject to change, especially as a result of human creativity.

In the context of failed states, the above approach leads to the issue of relations between endogenous institutions emerging from local knowledge and formal institutions in particular state structures. The functioning of the latter is not possible in isolation from the former (Platteau 2000; Aoki 2001). This means that the ability of institutions to persist and function according to the role assigned to them requires their convergence with informal institutions (Boettke et al. 2008). In this sense, they must be based on understandable and accepted values, norms and practices present in the community. The lack of such convergence limits the ability of planned institutions to achieve their intended goals. Moreover, while an institution may be effective in a particular area, this is not subject to simple extrapolation. The effectiveness of changes to socio-political structures depends on their embedding in local knowledge. So noticing and bearing in mind the role of endogenous structures is a prerequisite for the success of attempts to stabilise the situation in fragile states. As Boettke et al. (2008) show, an approach that takes into account the role of local knowledge can be used, among other things, to explain the successes and failures in the post-war reconstruction of some countries, as well as the fundamental differences in the effectiveness of the economic transition of former Eastern Bloc countries.

Noticing the role of local institutions leads to criticism of measures aimed at stabilising the situation by strengthening and increasing the effectiveness of centralised power structures. The theory of spontaneous order opposes state-centric approaches to the issue of failed states. Attempts to overcome dysfunctionality through such approaches lead to an undermining of the importance of local structures, and thus to the disruption of the built-in mechanisms of emergence and transmission of information. This leads to the obstructing of endogenous platforms for cooperation between agents. Centralisation of power also curtails the access of decision-makers to locally produced knowledge (Huerta de Soto 2010b, 52–62). Consequently, while reinforcing state structures (understood as centralisation of decision-making processes) is intended to strengthen cooperative tendencies, in the final analysis – together with weakening the importance of endogenous institutions – it may lead to an unintended decline in the attractiveness of cooperation within new structures. In particular, a change in the institutional context, by influencing the mechanisms of producing and transmitting information, may lead to an undermining of established practices and social norms. And the resulting increase in uncertainty promotes the growth of non-cooperative attitudes. In this sense, institutional changes may lead to an overall reduction in coordination capacity compared to the original level. In the context of game theory, this may be presented as a situation in which a cooperative game emerging from the consolidation of power structures is modified by other games played by individual agents on the basis of their local knowledge. In this sense, the whole system is a metagame consisting of many simultaneously played games (Tsebelis 1990).

The discoordination effect of top-down solutions not engrained in local customs, may be observed in the example of the influence of the colonial system on the Masai land management system (Blewett 1995). In the pre-colonial period, the system was based on common property and its use was regulated by unwritten social norms. The situation changed when the authorities replaced unwritten norms with formal contracts. The new system did not emerge from the codification of existing practices, but introduced solutions contrary to local knowledge on land management. This meant rejecting endogenous solutions. Consequently, it led to the undermining of the mechanisms of cooperative agriculture and to creating a conflict between the local actors, which manifested itself in the practice of rent-seeking.

While the ability of centralised power structures to perform the functions assigned to them depends on their relations with endogenous institutions, this does not mean that these structures cannot exist in the event of incompatibility with the said institutions. In such a case, however, they are not able to fully achieve their goals. The incompatibility of local institutions with the order created as part of the consolidation of power structures favours the emergence of conflict situations. And the resulting need for non-cooperative strategies increases the costs of implementation of tasks assigned to centralised structures, which makes their implementation impossible or limited (Autesserre 2010, 3; Coyne and Pellillo 2011, 5–8).

This does not rule out the existence of state structures in some limited extent, e.g., in the form of the so-called predatory state or vampire state, where power only serves to provide benefits to the ruling group (Ayittey 1999; Clapham 1996). This practice is promoted by the process of hierarchisation and concentration of power, characteristic of state structures. Consequently, power structures, perceived as interest groups, are characterised by a high degree of organisation and consolidation in comparison to the general public. And the centralised system of power limits the bottom-up ability of agents to coordinate their actions outside the state structures. All of this lowers the costs of actors pursuing their interests by operating within the power structures.[4] This involves these actors adopting non-cooperative strategies opposite other social structures.

Low compatibility with local endogenous institutions leads to attempts at using other sources of funding that enable the survival of power structures. This does not rule out the exploitation of local conditions (e.g., clan and kinship divisions) in the pursuance of self-interest. However, it does support the efforts of the power centre to create mechanisms reducing its dependence on these conditions. In particular, this leads to greater sensitivity to the expectations of external donors than to the preferences of local people (Eriksen 2011, 240). Combined with the limited access of foreign institutions to local knowledge, this produces problems with implementing "external" aid projects (Bueno de Mesquita and Downs 2006; Coyne 2007; Moyo 2010) and explains the negative impact of foreign aid on the condition of democratic institutions (Djankov et al. 2008). Moreover, it allows us to identify the cause and mechanism of the so-called resource curse in some countries. This term describes the observed slower economic growth of resource-rich countries compared to countries with scarce natural resources (Sachs and Warner 2001). Natural resources provide a fount of income largely independent of the condition of the economy in the country. Thus, in line with the approach presented here, they are an important factor enabling state power structures to survive regardless of their compatibility with local social institutions and the prosperity of the population.

An important aspect influencing the effects of reconstructing failed states is the participation requirement resulting from the territorial context of the state (see Kuniński 2004). This means that if structures working as a non-cooperative game are adopted, it is not possible to abandon the game and thus avoid the losses associated with possible defeat. And as a source of power, and at the same time a mechanism for redistributing goods, state structures are a form of leverage. Using them to pursue particular interests makes it possible to multiply the benefits by transferring the costs to others.

Both the impossibility of abandoning the game and the scale of benefits resulting from having an influence on power lead to an increase in the amount of resources that individual participants are ready to sacrifice in order to win. For both the potential profit and loss are greater than in a case where participants can withdraw from the game, thus reducing how much they may lose to others. Simultaneously, the willingness to incur significant costs increases the intensity of the conflict and the likelihood of the parties applying violent and costly solutions. Such promotion of non-cooperative behaviour indicates that the state itself may be a destabilising factor and may limit the ability of the bottom-up order, defined by the functions attributed to it, to emerge. This follows from the perception of the state as a zero-sum game, rather than serving the general public. Consequently, the dysfunctionality of the state may not be a manifestation of the absence of a (strong) state, but of the unreliability of this institution.[5]

The role of the state as a conflict-generating factor that promotes destabilisation of social structures is particularly evident in countries where decentralised clan and kinship structures and the customary law system play a major role in shaping the social order. It should be noted, however, that since local institutions, as a vehicle of social knowledge and practices, reflect informal, often inarticulate knowledge and practices resulting from experience, their impact on attempts to introduce institutional changes is general rather than limited to the cases mentioned above (see D'Amico 2012).

The example of Somalia

The negative effects of attempts to build centralised state structures are well illustrated by the example of Somalia. As concerns the issue of local knowledge and endogenous institutions, it is crucial to note that the structure of Somali society is based on the kinship system, which historically follows from the pastoral and nomadic culture represented by most Somalis (Lewis 1999). Although since the period of colonisation, processes such as urbanisation, the settling down of nomadic groups and changes in the economic structure have contributed to the emergence, mainly in cities, of new social groups and structures, and have weakened the importance of traditional structures, kinship bonds still remain an essential source of identity. Moreover, throughout the post-colonial period, the strong embedding of social relations in the kinship system has resulted in the absence of an effective central power exercising real control over the entire country. The kinship system is decentralised in its essence, which manifests itself in the political independence of individual groups, formed on the basis of clan affiliations and serving to protect their members. This means that in the absence of a central institution that can impose its decisions, relations between individual groups are based on consensual decision-making.[6] And the basic element shaping such a form of social order and enabling its functioning is the local customary law *Xeer* (Notten 2005). Such a law cannot be controlled and shaped from above, which promotes contractual relations between particular groups.[7]

The order based on kinship structures and customary law sharply contrasts with the order based on state institutions. The latter is characterised by the consolidation of power structures, including those related to legislation. The decision-making process, which in a decentralised system depended only on the interested parties, is hierarchised here, which limits the political independence of individual groups. Consequently, state structures are a source of political strength and benefit, so the participants desire to gain the largest possible influence on these structures. Importantly, however, the very creation or reconstruction of state institutions is combined with a strong embedding of social relations in kinship divisions. These divisions, as rooted in local norms and beliefs, have a significant impact on the perception of political processes and power struggles. Consequently, the pursuit and maintenance of power is naturally based on the exploitation of these divisions. This is particularly evident in African countries, where ethnic divisions still play an important role (Healy 2010, 373).

Due to the above, the possibilities of cooperation at the government level are limited, just as they are within clan structures. The consolidation of power and strengthening of state institutions stands in opposition to the egalitarianism of Somali society (Notten 2005). The traditions of this society, strongly shaped by the role of personal and group independence, do not recognise the division into the ruled and the rulers, and the process of political decision-making is based on consensus. Centralisation of power is therefore seen as undermining the idea of freedom. This erodes mutual trust and thus the ability to cooperate. The corresponding increase in uncertainty and the desire to secure the status quo and influence on the decision-making centre increases the attractiveness of non-cooperative behaviours. Voluntary cooperation is transformed into obligatory competition.

The decentralisation of socio-political structures not only promotes the perception of the state as a form of prey or reward and an instrument for the protection of the interests of a given group; it also hampers the emergence of a party system, characteristic for Western democracies, in which the essential element of the power struggle mechanism is party competition for the votes of the electorate. As a consequence, attempts to create or rebuild central structures in fragile states reduce trust between various actors and groups. This weakens their ability to cooperate and increases the attractiveness of non-cooperative and aggressive conduct in the securing of position. Political relations tend to be perceived as a "win-lose" situation, because gaining power opens the possibilities of using it to achieve political and economic benefits at the expense of others, while losing it poses the opposite risk. Participation in the power struggle is therefore important not only because of the possible benefits, but also, and perhaps more importantly, because of the threat to an actor resulting from subordination to the authority of others.

The above approach can be described as akin to the Hobbesian concept of a war of all against all. However, unlike in Hobbes, this is not due to the absence of the state, but to the desire to control it. Consequently, it links conflict with the existence of state structures, or the prospect of their creation. This approach is corroborated by the events that took place in Somalia after the fall of Siad Barre's regime in 1991 and the ultimate collapse of state structures it resulted in. This was the culmination of internal fighting since 1988, which had not produced the emergence of a new government. Along with the famine in the region brought about by political chaos and drought, this led to the UN intervention of 1992. During the operation, initially planned as a peace mission, the involvement of the international community was transformed into an attempt to rebuild the structures of the state. However, when stopping the fighting proved impossible the mission was terminated in 1995. But after the withdrawal of the international forces the intensity of the conflict decreased significantly. The situation was similar with subsequent attempts to establish central institutions. The prospect of talks, supported by the international community, produced the desire of individual actors to achieve the best possible starting position before the negotiations. They saw an opportunity to achieve a large impact on the government structures which were expected to result from these talks. And this led to an outbreak of fighting, also in previously peaceful regions. Conflicts appeared not only between clans, but also inside them. The need to appoint representatives to the talks sparked attempts by individual actors to gain a leadership role within the clan structures (Coyne 2006, 350).

Pointing at the discoordinating role of the state may also serve to explain the systems of patronage emerging in failed states and the corresponding phenomenon of predatory states. This was also the case when the Somali state still existed and served the interests of a narrow group of rulers, especially through using kinship affiliation as an effective means of exercising power (Lewis 2002, 250–257).

Implications

The theory of spontaneous order, pointing as it does to the issue of knowledge and the role of endogenous institutions, rejects the state-centric view of the dysfunctionality of the state. The state-centric approach is based on the existence of the state as a necessary condition for the implementation of the functions assigned to it, assuming also its overall ability to achieve that goal. This constitutes the normative aspect of the concept of state dysfunctionality, promoting certain forms of organisation of the power system's structures (Patrick 2011, 21–22; Call 2010, 303–304). Such a perspective produces a pronounced interventionist strand widely present in reflections on fragile states (Rotberg 2003; François and Sud 2006; Di John 2010), assuming the possibility of shaping social and political structures in accordance with the goals adopted by the planners. This promotes the perception of dysfunctionality as a management problem (Fukuyama 2004). For the theory of spontaneous order, such an approach is a sign of the fatal conceit of reason (see Hayek 1988). It ignores both the limitations in the availability of knowledge needed for the deliberate formation of such structures and the importance of local conditions. This approach also leads to attempts at strengthening state institutions and expanding their ability to perform their assigned functions, thus increasing the state's potential to influence various spheres of social life (Boege et al. 2009b, 17).

Criticism coming from the spontaneous order theory also concerns approaches that pay more attention to the issue of endogenous institutions and non-state actors (Lund 2006; Merkhaus 2007; Malejacq 2016). Although these models notice the problem of the influence of local conditions on the formation of the institutional order of the state, the state-centric perspective is maintained. This is particularly evident in the concept of hybrid political orders. This approach perceives the dysfunctionality of the state as resulting from the presence of other actors with a strong political position in relation to the state in a given area (Boege et al. 2009b). Consequently, it is necessary to take these actors into account and involve them in the process of state-building. In this sense, the hybrid approach describes an order based on a combination of traditional local forms of governance with institutional forms transplanted from the Western model of statehood. Since the system is rooted in local conditions, it becomes possible to strengthen the position of the state by founding it on the legitimization of non-state actors and on their capacity to deliver public goods

(Boege et al. 2009a, 20). In such an order, the state is just one of many actors providing public goods (Clements et al. 2007, 48). Non-state actors and state authority become two sides of the same coin.

The above approach to the issue of fragile states leads to discarding the Western model of statehood as optimal for the execution of functions assigned to the state. This is a response to the failure of previous attempts to strengthen state institutions (Pospisil and Kühn 2016, 9). This model, characterised by the inclusiveness of the political system and based on the mechanism of power sharing, is used to describe only some of the discussed states. It seems inadequate for most developing and unstable states (Balthasar 2015, 27–28). Under the hybrid approach, their description must account for the role of non-state actors. Consequently, it formulates a more holistic and historical approach to state-building processes, drawing attention to the importance of the context in which they are located (Haldén 2013, 50). Grounding the system in local institutional conditions not only strengthens state structures, but also deepens democratic mechanisms (Logan 2009, 123–124).

Emphasising the role of local conditions, the hybrid model and similar approaches offer a valuable perspective for analysing state dysfunctionality and attempts to address it, but they also have major weaknesses. In particular, they are not very good for identifying the factors and mechanisms conducive to the creation of sustainable and stable forms of statehood within hybrid orders (Balthasar 2015, 28–29). This means that they are not able to identify the exact causes of dysfunctionality and effective ways to counteract it. This poses the risk of "absolutising" local institutions regardless of their compatibility with the vision of the state order being formed. This means that they can be both constructive and destructive for processes of state formation. Instead of legitimising state structures, non-state actors may contribute to their weakening. In particular, although they may support the process of forming central authorities, this does not guarantee that the political order created in this way will be sustainable. The dynamic nature of relations between the state and local institutions means that they can change over time and destabilise the political situation.

As Markus Hoehne (2013) points out, this can be observed in Somaliland, where after the initial success of building a hybrid political order, the importance of the state increased. This meant a shift in the political centre of gravity towards centralised power structures, which gained an advantage over kinship-based, decentralised forms of politics. The participation of representatives of local institutions in state structures, entailing greater power and influence, led to abuses and corruption. This resulted in their alienation and loss of legitimacy in the eyes of the community they represented. And this, in turn, weakened local institutions and reduced the importance of clan structures. And although state power was strengthened, its legitimacy was undermined and the process of democratisation slowed down (see Bradbury 2008). Examples from other countries also show ambiguous effects of attempts to use local structures to strengthen state institutions (Podder 2014; Simangan 2018).

The above situation makes it possible to identify an important controversy in the area of fragile states, namely the issue of the influence of non-state actors and local institutional conditions on the ability to build and manage state structures. On the one hand, some researchers who are close to the hybrid approach point to the use of local structures in the process of forming the state as a factor conducive to its efficiency (Boege et al. 2009a; Mallett 2010; Johnson and Hutchison 2013). On the other hand, such a solution may be counterproductive and curtail the effectiveness of the state. Moreover, local structures may be incompatible with the state order. In the context of state-building processes based on the Western model of democracy, it cannot be ruled out that local institutions will be repressive and non-liberal. It should also be noted that the contrast between the two approaches is reflected in different perceptions of the idea of statehood (Balthasar 2015). Adopting a liberal approach, based on the idea of pluralism, is a premise for accepting the concept of a hybrid political order, while focusing on the aspect of state unity, which expresses the need to standardise the structures of socio-political life, brings attention to the destructive implications of the hybrid approach.

Both of these research perspectives seem to accurately identify mutual weaknesses. This allows presenting the issue of state dysfunctionality as a problem of finding the right balance between the legitimacy of the state and its efficiency (Call 2008a; Krasner and Risse 2014). However, this problem is considered from the state-centric perspective and does not undermine the necessity of the state's existence. While the differing approaches take into account the importance of local conditions, the point of reference for them is the impact of these conditions on the functioning of state structures. In particular, approaches suggesting the need to form hybrid political orders also treat them as instruments for the formation of the state order.

From the perspective of the spontaneous order theory, this approach seems incomplete. It is based on the assumption of the fundamental role of the state for coordination and ensuring social order, which means that what is under consideration is the problem of the effectiveness of the state. Thus, it ignores the issue of whether the state's assumption of these roles is legitimate. In other words, it focuses on the question of how to consolidate decision-making processes, bypassing the issue of the legitimacy of this consolidation.

The dispersion of knowledge and its role in the formation of social relations clearly indicates the importance of local knowledge for coordination processes. This does not mean, however, that the incorporation of endogenous institutions, which constitute the basis for this knowledge, in the mechanism of a centralised decision-making system solves the problem of lack of information needed to make decisions within this system. This problem results from the very fact that the decision-making process is centralised and that information is therefore difficult to obtain. Moreover, the formation of centralised decision-making structures involves the interference in the system of social institutions serving to create, transmit and use dispersed local knowledge. Going back to our previous reflections, this means that focusing on the effectiveness of the state reduces the problem to the identification of the actors participating in the decision-making process, while ignoring the issue of their knowledge.[8] Although this approach allows for the identification of local structures, it bypasses the question of the mechanisms of knowledge-formation within them. As such, it leads to the denial of the coordinating role of endogenous institutions, which express informal and often inarticulate knowledge and practices resulting from experience. Thus, it undermines the order created by these institutions and is a source of conflict. In contrast to the hybrid approach, this not so much suggests the need to take into account the situational context when forming the institutions of the state, but indicates that this context curtails the possibility of forming order according to the functions assigned to the state.

The issue of the central decision-making body having no access to information emerging in the course of social processes seems particularly problematic in the area of highly decentralised and dynamic social structures, characterised by a complex system of interdependencies. These features are particularly noticeable in some kinship systems (Notten 2005). The lack of necessary knowledge may manifest itself already at the stage of selecting participants in the state-building process and members of decision-making bodies. In an externally imposed process of building state structures, they are identified by external initiators of this process and the selection does not have to result directly from the actual role of the chosen representatives in local socio-political structures. In particular, representatives without real legitimacy from local communities may be selected. This problem was evident, for example, during two conferences aimed at forming a Somali government - – in Arthra in 2000 and in Mbagathi in 2002. Sixty per cent of those invited for the first conference were members of parliament from the Siad Barre's dictatorship. In an effort to strengthen the political stature of the second conference, the organisers invited not only representatives of local clans, but also major militia leaders and warlords. The selection of participants was also questioned because of the fact that many delegates were unable to safely return to the areas officially represented by them (Lewis 2008, 82–83). It should also be noted that the very formation of the central decision-making body makes it impossible to take into account the dynamic nature of kinship-based structures. Especially in the context of individual groups' aspirations to have their own representation, this may lead to changes in these structures and to an increase in the overall number of clans (Little 2003, 47).

Regardless of the issue of availability of knowledge, the occurrence of spontaneous order formation points to the potential possibility of the bottom-up and state-independent implementation of the functions assigned to government structures. This suggests that the issue of state dysfunctionality should not be reduced to the effectiveness of decision-making mechanisms within a centralised system of power, but should be based on a broader question of the legitimacy of consolidating decision-making processes. Examination of the problem of counteracting dysfunctionality cannot, therefore, be limited to the possibility of introducing democracy in a given country and should also look at the question of whether creating the possibility of making deliberative choices in a given area is justified. The theory of spontaneous order draws attention to the fact that order does not have to result from a conscious decision aimed at its creation. As regards the key role of law for the functioning of the social order, this does not rule out implementation of the legal order outside the institution of the state and the statute law system based on it. Such a system may operate not only on the basis of kinship affiliations, but also on the basis of religious communities (Bernstein 1992), functional links (e.g., commercial, Benson 1989), geographical proximity (Benson 2011, 21–30), reputation mechanism (Ellickson 1991) or exclusion (Stringham 2003).

The above approach undermines not only the legitimacy of attempts – manifesting the nirvana fallacy – to model fragile states on an idealised concept of the state or the relatively efficient Western countries (Coyne 2006). It is also a mistake to constrain the analysis to the issue of adapting state structures to local cultural conditions, as it does not question the very assumption of the necessity for a centralised decision-making system to exist (cf. Notshulwana 2011; Ghani and Lockhart 2008). As Peter Leeson and Claudia Williamson (2009) point out, in many cases replacing a fragile state with a functional one is not possible. In normative terms, this means that the best alternative for some countries may be anarchy. It cannot be ruled out that the absence of a state may foster cooperation and social order to a larger extent than inefficient state structures that weaken local institutions. The current global predominance of the institution of the state does not rule out alternative solutions (Benson 2011; Solvason 1993; Powell et al. 2008; Peden 1977)[9]. Moreover, as Martin Van Creveld (1999) points out, the state in its present form, based on the ideas of sovereignty and nation, is a relatively recent invention. Historically it is not the only form of socio-political order that emerged from the feudal system during the transformation of the medieval economy (Tilly 1992; Spruyt 1996).

The economic situation of Somalia after the fall of Siad Barre's regime in 1991 provides an example illustrating the validity of departing from the state-centric approach. The complete disappearance of state structures did not cause any deterioration in the economic situation. On the contrary, there was a significant increase in market activity and a major improvement in the condition of the economy compared to the period before 1991 (Mubarak 1997; Leeson 2007a; Powell et al. 2008). The reasons include the collapse of the Somali economy under the Siad Barre dictatorship, associated with an attempt to introduce a socialist economic model (see Huerta de Soto 2010c, 49–98), as well as the predatory practices of the state apparatus, which was only used to provide benefits to those in power. This meant the impotence of the state to perform control functions over the economy. Therefore, the lack of state regulation and control over the market after 1991 did not have a negative impact on the economy. Economic relations were based on customary law, which had informally performed the regulatory function also for the unofficial economy when the state still existed (Little 2003). All this made economic recovery after the collapse possible. It is worth noting that the situation improved not only in comparison to the previous period, but also as opposed to other countries of the region with relatively efficient state structures (Leeson 2007a).

In the context of the links between economic development and the stability of the social order, the above example can be seen as corroborating the hypothesis that the existence of a stable order is generally not conditioned by the presence of centralised decision-making structures, but by the level of uncertainty and risk, which is ambiguously correlated with the issue of statehood. This example also suggests that state structures themselves may be the cause of state dysfunctionality.

Rejecting the state-centric approach, the spontaneous order theory offers a broad perspective on the problem of fragile states and relates the issue to the whole spectrum of social phenomena associated with the instability of social, political, legal and economic structures. Rejecting the approach that sees the state as an institution determining the existence of social and legal order, it treats it as a factor that interferes with other institutions operating in a given area. In this sense, the state is just one of many institutions shaping this order. And state failure leads the theory to question the approach identifying factors determining state dysfunctionality with phenomena that weaken state-forming processes.

Notes

1 Other terms used to describe this phenomenon include: failed state, weak state or collapsed state. For terminology see Call (2008b), Kłosowicz (2013), Ścigaj (2013, 42–45), Pospisil and Kühn (2016).
2 Such an approach indicates the practical aspect of the notion of a state, whose meaning, and thus the range of functions considered to be within the powers of the state, is subject to change (see Eriksen 2010, 29–33).
3 The idea of local knowledge is close to the concept of *mētis* formulated by James Scott (1998, 6–7), concerning knowledge engrained in social structures (see Coyne and Boettke 2006; Pojevich 2003; Boettke et al. 2008).
4 It is worth noting that the above approach to predatory states can be applied to all centralised power structures. It helps explain the causes of domination of monocentric political systems in the contemporary world. Cf. Spruyt 1996.
5 See also Section "Implications" of Chapter 6.
6 It is worth noting that, unlike in other regions of Africa, most clans do not have the institution of a chief or leader. Power is exercised by the elders. This is a strong manifestation of independence, both in terms of the individual and the community (Notten 2005).
7 As a consequence, this social order is sometimes described as kritarchy or kritocracy (Dun 2005). This term, meaning "rule of the judges", stems from the role of judges in resolving disputes and the importance of the juridical processes in the formation of social order.
8 See Sections "Market processes and political processes" and "Democracy" of Chapter 7.
9 For an overview of the available literature on the stateless order and the processes of replacing anarchistic systems with the institution of the state see Powell and Stringham

2009. The issue of forming the legal order outside the state is also addressed in Section "Market processes and political processes" of Chapter 7.

Bibliography

Aoki, Masahiko. 2001. *Toward a Comaprative Instiutional Analysis*. Cambridge, MA: MIT Press.

Autesserre, Séverine. 2010. *The Trouble with the Congo: Local Violence and the Failure of International Peacebuilding*. Cambridge: Cambridge University Press.

Ayittey, George B. N. 1999. *Africa in Chaos*. New York: Palgrave Macmillan.

Balthasar, Dominik. 2015. From Hybridity to Standardization: Rethinking State Making in Contexts of Fragility. *Journal of Intervention and Statebuilding* 9 (1): 26–47. doi: 10.1080/17502977.2015.993502.

Benson, Bruce L. 1989. The Spontaneous Evolution of Commercial Law. *Southern Economics Journal* 55 (3): 644–661. doi: 10.2307/1059579.

Benson, Bruce L. 2011. *The Entreprise of Law: Justice without the State*. Oakland: The Independent Institute.

Bernstein, Lisa. 1992. Opting out of the Legal System: Extralegal Contractual Relations in the Diamond Industry. *Journal of Legal Studies* 21 (1): 115–157. doi: 10.1086/467902.

Blewett, Robert A. 1995. Property Rights as a Cause of the Tragedy of the Commons: Institutional Change and the Pastoral Maasai of Kenya. *Eastern Economic Journal* 21 (4): 477–490.

Boege, V., A. Brown, and K. Clements. 2009a. Hybrid Political Orders, Not Fragile States. *Peace Review* 21 (1): 13–21. doi: 10.1080/10402650802689997.

Boege, Volker, Anne Brown, Kevin Clements, and Anna Nolan. 2009b. *On Hybrid Political Orders and Emerging States: State Formation in the Context of "Fragility"*. Berlin: Berghof Research Centre for Constructive Conflict Management.

Boettke, Peter J., Christopher J. Coyne, and Peter T. Leeson. 2008. Institutional Stickiness and the New Development Economics. *American Journal of Economics and Sociology* 67 (2): 331–358. doi: 10.1111/j.1536-7150.2008.00573.x.

Bradbury, Mark. 2008. *Becoming Somaliland*. London: Progressio.

Bueno de Mesquita, Bruce, and George W. Downs. 2006. Intervention and Democracy. *International Organization* 60 (3): 627–649. doi: 10.1017/S0020818306060206.

Call, Charles T. 2008a. Building States to Build Peace? A Critical Analysis. *Journal of Peacebuilding and Development* 4 (2): 60–74. doi: 10.1080/15423166.2008.395667984152.

Call, Charles T. 2008b. The Fallacy of the "Failed State". *Third World Quarterly* 29 (8): 1491–1507. doi: 10.1080/01436590802544207.

Call, Charles T. 2010. Beyond the 'Failed State': Toward Conceptual Alternatives. *European Journalof International Relations* 17 (2): 303–326. doi: 10.1177/1354066109353137.

Clapham, Christopher. 1996. *Africa and the International System: The Politics of State Survival*. Cambridge: Cambridge University Press.

Clements, Kevin, Volker Boege, Anne Brown, Wendy Foley, and Anna Nolan. 2007. State Building Reconsidered: The Role of Hybridity in the Formation of Political Order. *Political Science* 59 (1): 45–56. doi: 10.1177/003231870705900106.

Coyne, Christopher. 2006. Reconstructing Weak and Failed States: Foreign Intervention and the Nirvana Fallacy. *Foreign Policy Analysis* 2 (4): 343–360. doi: 10.1111/j.1743-8594.2006.00035.x.

Coyne, Christopher. 2007. *After War: The Political Economy of Exporting Democracy*. Stanford: Stanford University Press.

Coyne, Christopher, and Peter J. Boettke. 2006. The Role of the Economist in Economic Development. *The Quarterly Journal of Austrian Economics* 9 (2): 47–68. doi: 10.1007/s12113-006-1008-y.

Coyne, Christopher, and Adam Pellillo. 2011. The Art of Seeing Like a State: State Building in Afghanistan, the DR Congo, and Beyond. *The Review of Austrian Economics* 25 (1): 35–52. doi: 10.1007/s11138-011-0150-8.

Creveld, Martin Van. 1999. *The Rise and Decline of the State*. Cambridge: Cambridge University Press.

D'Amico, Daniel J. 2012. Comparative Political Economy When Anarchism is on the Table. *The Review of Austrian Economics* 25 (1): 63–75. doi: 10.1007/s11138-011-0169-x.

Di John, Jonathan. 2010. The Concept, Causes and Consequences of Failed States: A Critical Review of the Literature and Agenda for Research with Specific Reference to Sub-Saharan Africa. *The European Journal of Development Research* 22 (1): 10–30. doi: 10.1057/ejdr.2009.44.

Djankov, Simeon, Jose G. Montalvo, and Marta Reynal-Querol. 2008. The Curse of Aid. *Journal of Economic Growth* 13 (3): 169–194. doi: 10.1007/s10887-008-9032-8.

Dun, Frank Van. 2005. Kritarchy. In *The Law of the Somalis: A Stable Foundation for Economic Development in the Horn of Africa*, edited by Spencer Heath MacCallum, 187–196. Trenton, NJ: The Red Sea Press.

Ellickson, Robert C. 1991. *Order without Law: How Neighbors Settle Disputes*. Cambridge: Harvard University Press.

Eriksen, Stein. 2010. The Theory of Failure and the Failure of Theory: "State Failure", the Idea of the State and the Practice of State Building. In *Troubled Regions and Failing States: The Clustering and Contagion of Armed Conflicts*. Vol. 27 of *Comparative Social Research*, edited by Kristian Harpviken, 27–50. Bingley, UK: Emerald Group Publishing.

Eriksen, Stein. 2011. "State Failure" in Theory and Practice: The Idea of the State and the Contradictions of State Formation. *Review of International Studies* 37 (1): 229–247. doi: 10.1017/S0260210510000409.

François, Monika, and Inder Sud. 2006. Promoting Stability and Development in Fragile and Failed States. *Development Policy Review* 24 (2), s. 141–160. doi: 10.1111/j.1467-7679.2006.00319.x.

Fukuyama, Francis. 2004. *State-Building: Governance and World Order in the 21st Century*. Ithaca: Cornell University Press.

Ghani, Ashraf, and Clare Lockhart. 2008. *Fixing Failed States: A Framework for Rebuilding a Fractured World*. Oxford: Oxford University Press.

Haldén, Peter. 2013. Against Endogeneity: The Systemic Preconditions of State Formation. In *New Agendas in Statebuilding – Hybridity, Contingency and History*, edited by Robert Egnell, and Peter Haldén, 32–54. London, New York: Routledge.

Hayek, Friedrich A. 1988. *The Fatal Conceit: The Errors of Socialism*. London: Routledge.

Healy, Sally. 2010. Reflection on the Somali State: What Went Wrong and Why it might not Matter. In *Peace and Milk, Drought and War: Somali Culture, Society and Politics*, edited by Markus Hoehne, and Virginia Luling, 367–384. London: Hurst Publishers.

Helman, Gerald B., and Steven R. Ratner. 1992. Saving Failed States. *Foreign Policy* 89: 3–18. doi: 10.2307/1149070.

Herbst, Jeffrey. 1997. Responding to State Failure in Africa. *International Security* 21 (3): 120–144. doi: 10.2307/2539275.

Hoehne, Markus V. 2013. Limits of Hybrid Political Orders: The Case of Somaliland. *Journal of Eastern African Studies* 7 (2): 199–217. doi: 10.1080/17531055.2013.776279.

Huerta de Soto, Jesus. 2010b. *The Theory of Dynamic Efficiency*. London: Routledge.

Huerta de Soto, Jesus. 2010c. *Socialism, Economic Calculation and Entrepreneurship*. Cheltenham, UK: Edward Elgar Publishing.

Jackson, Robert H. 1990. *Quasi-States: Sovereignty, International Relations and the Third World*. Cambridge: Cambridge University Press.

Johnson, Kristin, and Marc L. Hutchison. 2013. Hybridity, Political Order and Legitimacy: Examples from Nigeria. *Journal of Peacebuilding and Development* 13: 37–52. doi: 10.1080/15423166.2012.743811.

Kaplan, Seth D. 2008. *Fixing Fragile States – A New Paradigm for Development*. Westport: Praeger Security International.

Kłosowicz, Robert. 2013. Wstęp. In *Państwa dysfunkcyjne i ich destabilizujący wpływ na stosunki międzynarodowe*, edited by Robert Kłosowicz, 7–9. Kraków: Wydawnictwo Uniwersytetu Jagiellońskiego.

Krasner, Stephen D., and Thomas Risse. 2014. External Actors, State-Building, and Service Provision in Areas of Limited Statehood: Introduction. *Governance* 27 (4): 545–567.

Kuniński, Miłowit. 2004. Państwo. In *Słownik społeczny*, edited by Bogdan Szlachta, 801–824. Kraków: Wydawnictwo WAM.

Leeson, Peter T. 2007a. Better off Stateless: Somalia Before and After Government Collapse. *Journal of Comparative Economics* 35: 689–710. doi: 10.1016/j.jce.2007.10.001.

Leeson, Peter T., and Claudia R. Williamson. 2009. Anarchy and Development: An Application of the Theory of Second Best. *The Law and Development Review* 2 (1): 76–96. doi: 10.2202/1943-3867.1032.

Lewis, Ioan M. 1999. *A Pastoral Democracy: A Study of Pastoralism and Politics Among the Northern Somali of the Horn of Africa*. Hamburg: LIT Verlag.

Lewis, Ioan M. 2002. *A Modern History of the Somali: Nation and State in the Horn of Africa*. Athens: Ohio University Press.

Lewis, Ioan M. 2008. *Understanding Somalia and Somaliland*. London: Hurst.

Little, Peter D. 2003. *Somalia: Economy without State*. Oxford: James Currey.

Logan, Carolyn. 2009. Selected Chiefs, Elected Councillors and Hybrid Democrats: Popular Perspectives on the Co-Existence of Democracy and Traditional Authority. *Journal of Modern African Studies* 47 (1): 101–128. doi: 10.1017/S0022278X08003674.

Lund, Christian. 2006. Twilight Institutions: Public Authority and Local Politics in Africa. *Development and Change* 37 (4): 685–705. doi: 10.1111/j.1467-7660.2006.00497.x.

Malejacq, Romain. 2016. Warlords, Intervention, and State Consolidation: A Typology of Political Orders in Weak and Failed States. *Security Studies* 25 (1): 85–110. doi: 10.1080/09636412.2016.1134191.

Mallett, Richard. 2010. Beyond Failed States and Ungoverned Spaces: Hybrid Political Orders in the Post-Conflict Landscape. *eSharp*, no. 15: 65–91.

Merkhaus, Ken. 2007. Governance without Government in Somalia: Spoilers, State Building, and the Politics of Coping. *International Security* 31 (3): 740–106. doi: 10.1162/isec.2007.31.3.74.

Moyo, Dambisa. 2010. *Dead Aid: Why Aid is Not Working and How There is Another Way for Africa.* London: Penguin Books.
Mubarak, Jamil A. 1997. The "Hidden Hand" Behind the Resilience of the Stateless Economy of Somalia. *World Development* 25 (12): 2027–2041. doi: 10.1016/S0305-750X(97)00104-6.
Notshulwana, Mxolisi. 2011. State Fragility in Africa: Methods Chasing Problems or Problems Chasing Methods in Political Discourse? *International Journal of African Renaissance Studies* 6 (2): 81–99. doi: 10.1080/18186874.2011.650850.
Notten, Michael van. 2005. *The Law of the Somalis: A Stable Foundation for Economic Development in the Horn of Africa.* Edited by Spencer Heath MacCallum. Trenton: The Red Sea Press.
Patrick, Stewart. 2011. *Weak Links: Fragile States, Global Threats, and International Security.* Oxford: Oxford University Press.
Peden, Joseph R. 1977. Property Rights in Celtic Irish Law. *Journal of Libertarian Studies* 1 (2): 81–95.
Platteau, Jean-Philippe. 2000. *Institutions, Social Norms, and Economic Development.* Amsterdam: Harwood Academic Publishers.
Podder, Sukanya. 2014. State Building and the Non-State: Debating Key Dilemmas. *Third World Quarterly* 35 (9): 1615–1635. doi: 10.1080/01436597.2014.970864.
Pojevich, S. 2003. Understanding the Transaction Costs of Transition: It's the Culture, Stupid. *Review of Austrian Economics*, 16 (4): 347–361. doi: 10.1023/A:1027397122301.
Pospisil, Jan, and Florian P. Kühn. 2016. The Resilient State: New Regulatory Modes in International Approaches to State Building? *Third World Quarterly* 37 (1): 1–16. doi: 10.1080/01436597.2015.1086637.
Powell, Benjamin, Ryan Ford, and Alex Nowrasteh. 2008. Somalia After State Collapse: Chaos or Improvement? *Journal of Economic Behavior and Organization* 67 (3/4): 657–670. doi: 10.1016/j.jebo.2008.04.008.
Powell, Benjamin, and Edward P. Stringham. 2009. Public Choice and the Economic Analysis of Anarchy: A Survey. *Public Choice* 140 (3–4): 503–538. doi: 10.1007/s11127-009-9407-1.
Rotberg, Robert I. 2002. Failed States in a World of Terror. *Foreign Affairs* 81 (4): 127–140.
Rotberg, Robert I. 2003. *When States Fail: Causes and Consequences.* Princeton: Princeton University Press.
Sachs, Jeffrey. D., and Andrew M. Warner. 2001. The Curse of Natural Resources. *European Economic Review* 45 (4–6): 827–838. doi: 10.1016/S0014-2921(01)00125-8.
Scott, James C. 1998. *Seeing Like a State.* New Haven: Yale University Press.
Simangan, Dahlia. 2018. Domino Effect of Negative Hybrid Peace in Kosovo's Peacebuilding. *Journal of Intervention and Statebuilding* 12 (1): 120–141. doi: 10.1080/17502977.2018.1423772.
Solvason, Birgir. T. R. 1993. Institutional Evolution in the Icelandic Commonwealth. *Constitutional Political Economy* 4 (1): 97–125.
Spruyt, Hendrik. 1996. *The Sovereign State and Its Competitors: An Analysis of Systems Change.* Princeton: Princeton University Press.
Stringham, Edward P. 2003. The Extralegal Development of Securities Trading in Seventeenth Century Amsterdam. *Quarterly Review of Economics and Finance* 43 (2): 321–344. doi: 10.1016/S1062-9769(02)00153-9.
Ścigaj, Paweł. 2013. Problemy metodologiczne w badaniu państw dysfunkcyjnych. In *Państwa dysfunkcyjne i ich destabilizujący wpływ na stosunki międzynarodowe*, edited by Robert Kłosowicz, 37–69. Kraków: Wydawnictwo Uniwersytetu Jagiellońskiego.

Tilly, Charles. 1992. *Coercion, Capital, and European States, AD 990–1992*. Oxford: Blackwell.
Tsebelis, George. 1990. *Nested Games: Rational Choice in Comparative Politics*. Berkeley: University of California Press.
Zartman, I. William. 1995. *Collapsed States: The Disintegration and Restoration of Legitimate Authority*. Boulder: Lynne Rienner.

Part IV

Criticism and limitations of the theory of spontaneous order

9 Limitations of the theory of spontaneous order

Limitations on the application of the theory of spontaneous order in political science

The theory of spontaneous order is based on a praxeological approach, manifested in the interpretation of social reality in terms of human action. And the prerequisite for human action is the existence of a state of uncertainty, meaning that the person taking action does not have full knowledge guaranteeing the achievement of intended goals. Consequently, we may present human activity as resulting from the ability to make authentic choices based on preferences. This also emphasises the creative nature of human conduct, linked to cognitive limitations.[1]

The above approach turns the issue of the availability of knowledge into the starting point for analyses within the spontaneous order theory. It implies an incompleteness and dispersion of knowledge possessed by agents and the impossibility of fully aggregating and articulating it. Since knowledge determines the actions of individuals and thus the social phenomena resulting from them, this theory denies human omnipotence in designing social reality. And the state of uncertainty leads to unforeseen effects of human activity, allowing for the emergence of spontaneous social structures.

As far as social phenomena are concerned, the occurrence of order determines the possibility of mutual coordination between individual agents. In the case of spontaneous order, coordination does not result from the planned pursuit of this order and deliberate formulation of specific rules of its functioning. In other words, its emergence is not the result of a plan based on a conscious decision. In the context of human creativity, this allows us to perceive social phenomena as a dynamic and open process of producing new information during individual interaction with the surrounding physical and social environment.

The unplanned nature of spontaneous order means that it is not intentional. It cannot be assigned a goal that would stand behind its emergence. The only recognisable goals are those of the individuals producing this order. In this sense, the functions of such an order are conceived as mechanisms enabling individuals to achieve their particular goals.

As based on the concept of human action and not constrained to a specific area of social reality, the spontaneous order theory embraces in its research field all purposeful human activity which can be said to have unintended effects. In this sense, the theory is general, and not limited to selected social phenomena. Due to the subjectivity and dispersion of knowledge, in all areas of human activity the pursuit of intended goals and coordination of activities take place in a state of incomplete information. So in terms of our theory, it is the situation of permanent uncertainty that defines the basic research issue of social sciences as such. The problem for these sciences is not phenomena resulting from decisions made on the basis of the complete knowledge needed to achieve a goal, but actions founded on knowledge that is not given to anyone in its entirety.

In view of this, the fundamental issue for social science is explaining the emergence and functioning of spontaneous order in the social world. This means trying to answer the question of how it is possible for social structures not resulting from deliberate decisions aimed at their establishment to emerge. To this end, the spontaneous order theory presents unplanned order as resulting from the process of the adaptation of individuals to local circumstances and from mutual coordination. This coordination is not based on the process of knowledge aggregation, but on the transmission of information as part of a dynamic process of interaction between individual agents, enabling the use of the dispersed knowledge possessed by particular individuals.

Assuming the existence of human cognitive limitations defines another area of research within spontaneous order theory. These limitations refute human omnipotence in the planned coordination of activities. This is particularly evident in the context of authentic human creativity and the tacit nature of part of people's knowledge. This provides the starting point for reflection on the limits of any deliberate shaping of social reality.

The spontaneous order theory thus expressed focuses on phenomena relating to spontaneous social structures. This implies limitations on its applicability. In particular, these restrictions relate to the area of political science and result from the characteristics of the subject of research.

The subject of political science research is multi-faceted and syndromatic, and as such cannot be implanted into a rigid framework.[2] Politics interpenetrates with other aspects of social life, which makes it impossible to separate a homogenous class of political phenomena. Therefore, defining the boundaries of what is political is always arbitrary and fluid. As Mirosław Karwat (2010, 78) points out, this is connected with the variability of social phenomena, which gives the subject of study an accidental and historically changeable character.[3] And the role of the (variable) context emphasises the importance of temporality as an important property of social sciences in general (Blok and Kołodziejczak 2015, 24). This is noticeable already at the level of understanding the notions of "politics" or "politicality" themselves, their historical character precluding the adoption of widely accepted definitions.

The impossibility of defining a unique object of research means that it does not form a calculable set of elements, but a fuzzy set without a precise and unambiguous criterion of belonging. Consequently, determining the subject of research is an intentional procedure (Pierzchalski 2013, 35). This makes the subject of research dependent on the person examining it. Its determination is an intentional procedure with its source in the cognising agent. This means that whether something belongs to the research area is based on individual decisions of the researcher, which implies that the discipline is non-homogenous.

Despite being indeterminate and fluid, political phenomena do have a distinguishing trait, namely the strongly teleological character of the social relations that constitute them. This means that, as concerns the form of social order, the research field of political science seems to focus on structures organised in a planned manner.[4] This is particularly evident in historical forms of perceiving politics, which were closely connected with the institution of the state. In this sense, the traditional approach to politics strongly correlates with the deliberate shaping of social relations through state power (Almond 1998, 50–96). The key role of the state is particularly noticeable in the institutional approach, which dominated in the initial period of development of political science, was important for the identity of this discipline, and focused on the issue of state power and the institutional mechanisms of its exercise (Rhodes et al. 2006, xii–xvii).

The above approach is present in the concept of politics conceived as the art of governance. In this perspective, typical for such thinkers as Niccolò Machiavelli or Weber, politics is identified with the pursuit of power and maintenance of it. Since political phenomena take place within the framework of the state, research on them concentrates on the functioning of state structures. Another example of the traditional understanding of politics is its identification with the public sphere and governing for the common good. Although this perspective does not refer directly to the state, it assumes the teleological character of the concept of the common good. Consequently, the state is here an institution serving the organisation of social life. This means that it has the public character of an instrument for achieving the goals of a given community.

Today we can notice the process of expanding the space of political phenomena and so of the area of research (Blok and Kołodziejczak 2015, 23–24). On the one hand, this naturally follows from the temporality of social science. On the other hand, it results from the syndromaticity of social phenomena, which in the context of the pluralism of political science makes it open to new issues, previously analysed by other disciplines. Consequently, politics is now regarded as a universal phenomenon going beyond the traditional concept of the state and placed in the broader context of the exerting of influence. This means a change in the relationship between the concepts of "politics" and "state". The state ceases to be perceived as a political monopoly, becoming just one of the actors of political life (Jabłoński 2012, 36). But political phenomena are now seen as universal also in the traditional context of state power, which is due to the expansion of the spectrum of government functions, and thus its significance and influence on various spheres of social life (Luhmann 1981).

It should be noted at this point that while the research space of political science is changing, this does not undermine the significance and role of historical approaches. Tradition strongly influences the landscape of the discipline. It is one of the factors shaping the identity of the researcher, serving in particular to justify the type of research undertaken (Ścigaj 2010, 39). And so it creates a framework which imposes limitations and guidelines on research but also facilitates cooperation (Goodin and Klingemann 1998, 5).

The expansion of the research area and changes in the perception of what is political make it possible to distinguish two approaches to understanding political phenomena (Blok 2009, 34). The first one is the subjective approach, close to the everyday understanding of politics. It identifies politics with a specific set of institutions, focusing especially on politics at the government level. In the second approach, political phenomena are more broadly seen as a social process that goes beyond the operations of the government. Within its framework, politics is implemented through the execution of power and dominance. This means politicising a wide range of aspects of social life. The two approaches correspond to one of the seven main axes of division distinguished by Bernard Grofman (1997) and adopted in political science. This creates a distinction between conceiving of politics as a sphere of governmental activity, and focusing on the issue of power and influence independently of the area in which they manifest themselves.

While the two approaches differ in their perceptions of political phenomena, they both focus on exploring planned and deliberate social relations and structures. Politics, understood as a type of social activity, is teleological. Its goal is to shape social order (Blok 2009, 42). In this sense, political activity is based on making decisions aimed at setting the goals that a given political community strives for. And the effect of this activity is the creation of a power structure (Klementewicz 2010, 62). Consequently, the deliberate shaping of the social order through political activity can also be seen in the relations between people involved in politics. In this perspective, the essence of politics is defined by the categories of consensus, conflict and interest (see Blok 2009, 48–54).

The wide spectrum of political phenomena forms an extensive network of subject areas of various research theories and traditions. Within this network, Klementewicz (2013, 32) distinguishes three main theoretical-methodological orientations. They differ in their understanding of politics, presenting different perceptions of the subject area. But all of them are characterised by both vitality and being deeply grounded in the research tradition.

In the first approach, politics is understood as making use of power, consisting in the establishment of legal norms and the use of legitimate physical coercion. Politics boils down here to "people's actions that consist in determining the sphere of application [...] of a means of regulating behaviour in society, in directing the way it is applied by legislating and enforcing the law" (Klementewicz 2010, 64–65). Consequently, a politician is seen in this perspective as an organiser and guarantor of social order.

In the second approach, politics is seen as a process of coordination and regulation of the social system through a broad range of instruments of power. The state is here a collective entity that acts to improve the functioning of society, while the politician is perceived as a helmsman of social order. This means that politics serves to turn social relations into a cooperative game and to increase the benefits following on from cooperation. Politics, therefore, makes it possible to achieve social goals that go beyond the provision of security (cf. Klementewicz 2010, 68–72).

In the third approach, the politician acts as a biased arbiter in conflicts over the distribution of wealth. This approach is based on pointing out conflicts and collisions of interest in the economic structure and perceiving politics as a reaction to them.

The three approaches focus on different aspects of politics. The first is concentrated on the main instrument of its pursuit, which is law. The subject of the second approach is the basic function of politics, namely the coordination of social structures. The third concerns the social sources and ramifications of politics. This multifacetedness of politics means that only the combination of all these perspectives may produce a comprehensive understanding of the political dimension of the social world (see Karwat 2009).

However, in all three approaches, one can see the perception of politics as an active and planned shaping of social order. They all describe it as expressed in decisions and actions aimed at transforming social structures. In the case of the first approach, this is due to the very fact of seeing political power as serving such a purpose. In the second, politics plays a coordinating role in the pursuit of specific social goals, which means that it is seen as a deliberate shaping of social structures. Consequently, the very existence of social order is conditioned by the possibility of its planned creation. And in the last perspective, the distribution of goods and achievement of goals is deliberately done by means of political power.

These observations suggest that politics is a "system of social relations of a teleological nature" (Laska 2014, 152). This means that political science focuses on issues related to the existence and functioning of planned social structures. This is an important limitation for the politological application of the spontaneous order theory, which focuses on explaining spontaneous phenomena and is rarely employed to explain the functioning of government institutions or state structures treated as deliberate creations.

In the context of the above discussion, the second of the research directions within the spontaneous order theory, dealing with limitations in the planned shaping of social order, seems much more important. By emphasising the cognitive limitations of agents and pointing at the crucial role of their knowledge in shaping social relations, this theory denies the possibility of human omnipotence in creating social order. In this sense, it poses a question about the limitations in achieving political goals of human actions. While a politician can be seen as an organiser of social order, the question of the extent to which he is capable of performing this function becomes important.

This raises the question of the role of law as an instrument of politics and its relations with the state. Although the distinguishing feature of politics is legitimate coercion in the organisation of social life, the question may be asked as to whether it is justified to understand such coercion as – by implication – issuing from the state (cf. Klementewicz 2010, 59). The spontaneous order theory draws attention to the question of whether it is valid to assume a monocentric character of the system of making and enforcing the law. In the context of cognitive limitations, this problem can be formulated as the question of whether the planned formation of a legal system that could achieve intended social goals is possible. Additionally, it raises the issue of alternative solutions in the form of polycentric legal orders.[5] The explanatory aspect of the theory under discussion once again grows in importance here, as it may serve to explain the process of the unplanned formation and functioning of such an order (see Long 2006; Boettke et al. 2008; Wiśniewski 2014a). In effect, these issues form the basis for reflection on the normative aspect of the legal order and the existence of the state (Nozick 1974; Long and Machan 2008; Casey 2012).

Limitations in the ability of humans to deliberately achieve goals stem not only from the existence of cognitive barriers. They can also be caused by the very system of planning and the top-down coordination of order. In the context of the coordinating and regulatory function of politics, this means that structures serving this function of themselves change the attractiveness of the solutions. In terms of game theory, this means that they change the distribution of winnings. In particular, this can affect the attractiveness of cooperative strategies. As the spontaneous order theory indicates, the existence of centralised decision-making structures limits the ability of agents to create and exchange information. This in turn limits the ability of players to discover new solutions (strategies) that increase the value of winnings. This can lead to a situation where losses associated with restricting players' access to information outweigh the benefits of the top-down coordination of their actions.

The negative impact of planned order shaping the flow of information between agents is particularly important in the context of the consensual understanding of politics. Deliberation and discussion serve the exchange of information. However, in the light of human creativity and the nature of the knowledge used in decision-making, this does not mean that actions initiated on the basis of a consensually defined goal are optimal. This understanding of politics focuses on the choice of goals and the mechanisms for achieving them. In contrast, the spontaneous order theory looks not so much at the way decisions are made, but at the problematic nature of subjecting a given issue to the decision-making process itself. This problem suggests the importance of the plane of interaction or competition between different interests, thus undermining the consensual approach to these processes.[6] This approach is based on considering these phenomena at the level of decisions aimed at consciously shaping them. In other words, it means seeing social interactions as planned. In contrast, while noting the existence of other planes, the spontaneous order theory allows us to pinpoint possible negative effects of such an approach to politics.

The spontaneous order theory seeks to explain unplanned structures, and this aspect of the theory does not produce a major added value for political science, which focuses on the study of the active and planned formation of social order. This concerns especially the narrow understanding of politics focusing on the government and the operation of state institutions. The above examples, regarding the issues of instruments, functions and the consensual aspect of politics, indicate that the problem of limitations in the purposeful shaping of social order, raised by our theory, seems to have much greater significance in the study of politics. In this respect, the theory allows attempts to explain the dysfunctionality of planned structures by invoking the issue of cognitive limitations.

Unclearness of the concept of spontaneous order as a problem situation

The theory of spontaneous order seems to provide a valuable approach to analysing the formation and change of social structures. However, distinguishing only two categories of order – spontaneous and planned – may give rise to doubts as to the extent to which such a dichotomous view corresponds to the wide spectrum of human interactions with all their complexity and as to the extent to which it allows for their explanation. This raises the question of both the precision of the definition of these categories and the validity of using a dichotomous perspective.

As Timothy Sandefur (2009, 5) notes, the distinction between spontaneous and planned order depends on the perspective of the observer. The absence of transparent criteria for this categorisation is well illustrated by the analyses of Richard Posner (2005) and Donald Boudreaux (2006) concerning the US Constitution. Posner suggests that the Constitution is an example of a deliberate project which effectively creates a planned order. But Boudreaux, who doesn't see the Constitution as establishing a new order, rejects such an interpretation. He believes that the Founding Fathers did not want to create all or even most laws from scratch, but based them on pre-existing common law. Moreover, this law, rooted in English history and modified by the experience of the colonies, remained the law of the land (*lex terrae*) and the Constitution altered them only to a moderate extent.

This may be seen as implying the need to clarify the definition or change the categorisation of phenomena. However, while precision of the terms used is undoubtedly an important element of scientific research, such an analysis should not be an end in itself, as Popper (2002) pointed out. Terms should be adapted to the research problem. Increases in knowledge result from identifying problems and attempting to solve them. In the context of research, terms are therefore only a means to an end and the degree of their accuracy should not go beyond what is needed in a given situation. According to Popper, the progress of science is not an outcome of terminological precision, but "of seeing new problems where none have been seen before, and of finding new ways of solving them" (2002, 23). Analyses of concepts and their meanings that go beyond this aim are described by him as verbalism; this not only wastes time and effort on useless deliberations, but also leads to a loss of clarity.

If the theory of spontaneous order is applied and the two basic forms of order (planned and spontaneous) formulated by it are distinguished from each other, the problem situation concerns the human limitations in shaping social reality. For although this theory serves to explain the origin of unplanned processes of coordination of individual actions, it does not define the limits of the human ability to create planned and spontaneous order. The key point here is to note the grounding of this theory in the assumption of human cognitive limitations. It follows that only with relatively simple structures can all the necessary information be aggregated in order to create them. In contrast, the creation of complex structures is only possible through dynamic processes enabling the use of the dispersed knowledge possessed by individual agents. Consequently, this theory combines the ability to form order with human cognitive limitations. An absence of the latter would mean rejecting the concept of action, and thus any classification of human interactions.

The above analysis indicates that the fundamental problem posed by the theory of spontaneous order is the question of human cognitive boundaries, as they determine the ability to shape social space. Consequently, the spontaneous and planned order distinction is only a reflection of the absence of human omniscience. According to the approach proposed by Popper, this does fulfil its role, allowing us to see the problem of indeterminate cognitive boundaries. At the same time, attempts to define conditions that would allow for a more precise identification of order within the complex sphere of human interaction are a form of verbalism (cf. Flori 2006; Sandefur 2009). This is particularly evident in Posner's and Boudreaux's investigations into the US Constitution. Their examinations of the question of whether the order founded on this legal act is spontaneous or planned concern definitions and as such are irrelevant. They do not contribute to our understanding of the subject, because they only focus on the pursuit of more precise terms, which boils down to verbalism.

The issue of cognitive boundaries is also present in the discussion on the normative aspect of spontaneous order, related to the concept of constitutional order formulated by Hayek (1960, 1979). Grounding the possibility to coordinate people's actions in individual freedom, Hayek pointed out that maximising this freedom requires the existence of a legal system based on general and impersonal rights. Only a constitutional system characterised by such universal norms guarantees the widest possible sphere of individual freedom of action. Only "when we obey laws, in the sense of general abstract rules laid down irrespective of their application to us, we are not subject to another human will and therefore free" (Hayek 1960, 153). This led Hayek to opt for institutional changes that would make this structure possible. However, according to some researchers (Barry 1994; Samuels 1999; Muller 2007), this attitude is inconsistent with Hayek's criticism of constructivist rationalism[7] and the pretence of knowledge. On the one hand, the limitations of individual knowledge pointed out by him made him sceptical about the planned construction of institutions. On the other hand, his concept of the constitutional order prompted him to make "a radical departure from established tradition" (Hayek 1973, 4) and support a reform of the constitutional system based on an examination of current beliefs and fundamental concepts, and a rational analysis of the faults of the democratic system. This led to the claim that two incompatible approaches are present in Hayek's thought (Kukathas 1989, 206).

Supporters of Hayek's concept of constitutional order (see Vanberg 1994; Steele 2002a; Servant 2013) reject such criticism, defending the possibility of a rational reform of the political system. They point out that the emergence of the extended order takes place within a certain framework that Hayek's proposed constitutional model is intended to provide. Although the evolutionary nature of social change implies human error, "To advocate the utilization of competitive evolutionary processes as a critical check on our institutional constructions is not the same as to claim that we could get along without any rational institutional design" (Vanberg 1994, 194). Thus it does not rule out deliberate creation of structures within this process. Just like the free-market competition, which is the source of bottom-up coordination processes, functions through the institution of property rights, political action should be analysed within certain institutional constraints.

However, our previous reflections suggest that this discussion does not in fact concern the alleged self-contradiction of Hayek's concept of constitutional order, but the limits of human cognition. They determine the possibilities of acquiring the knowledge needed to achieve a given goal. Knowing them is a necessary condition for determining whether a given regulation is not a manifestation of the fatal conceit of reason.

Notes

1 See Chapter 3.
2 See Chapter 5.
3 At the same time, the lack of this homogeneity makes it possible to perceive politicality as gradable (Karwat 2010, 78–81).
4 As an indicator, the highlighted feature does not aspire to being a criterion determining the essence of political phenomena. The teleological character is only their aspect. As has been signalled, this indicator refers to one of the aspects of the examined phenomena, namely the form (planned or spontaneous) of the social structures that make them up (see Karwat 2010, 66–67).
5 It is worth emphasising here the compatibility of such an approach with the perception of law in terms of legal realism presented by the theory of spontaneous order. Norms, including legal ones, are expressed in the practices and customs of people. In particular, this stands in opposition to the approach of legal positivism, which reduces the law to the norms of statutory law and thus establishes monocentricity

as a prerequisite for the existence of law. See also Section "The theory of spontaneous order in relation to the epistemic system of contemporary political science" of Chapter 5.

6 This issue is discussed in Chapter 7.

7 Hayek (1973, 5) defines constructivist rationalism as an approach – belonging to the Cartesian tradition – which says that all social institutions are, and should be, products of an intended plan – the pursuit of a top-down, planned construction of an order, coordinating the actions of individuals for a given purpose.

Bibliography

Almond, Gabriel A. 1998. Political Science: The History of the Discipline. In *A New Handbook of Political Science*, edited by Robert Goodin, and Hans-Dieter Klingemann, 50–96. Oxford: Oxford University Press.

Barry, Norman P. 1994. The Road to Freedom – Hayek's Social and Economic Philosophy. In *Hayek, Co-ordination and Evolution: Hayek, Co-ordination and Evolution: His Legacy in Philosophy, Politics, Economics and the History of Ideas*, edited by Jack Birner, and Rudy van Zijp, 178–189. London: Routledge.

Blok, Zbigniew. 2009. *O polityczności, polityce i politologii*. Poznań: Wydawnictwo Naukowe WNPiD UAM.

Blok, Zbigniew, and Małgorzata Kołodziejczak. 2015. O statusie i znaczeniu kategorii "polityki" i "polityczności" w nauce o polityce. *Studia politologiczne,* 37: 17–32.

Boettke, Peter J., Christopher J. Coyne, and Peter T. Leeson. 2008. Institutional Stickiness and the New Development Economics. *American Journal of Economics and Sociology* 67 (2): 331–358. doi: 10.1111/j.1536-7150.2008.00573.x.

Boudreaux, Donald J. 2006. Hayek's Relevance: A Comment on Richard A. Posner's, "Hayek, Law, and Cognition". *New York University Journal of Law and Liberty* 2 (1): 157–165.

Casey, Gerard. 2012. *Libertarian Anarchy: Against the State*. London: Continuum.

Flori, Stefano. 2006. The Emergence of Institutions in Hayek's Theory: Two Views or One? *Constitutional Political Economy* 17 (1): 49–61. doi: 10.1007/s10602-006-6793-y.

Goodin, Robert E., and Hans-Dieter Klingemann. 1998. *A New Handbook of Political Science*. Oxford: Oxford University Press.

Grofman, Bernard. 1997. Seven Durable Axes of Cleavage in Political Science. In *Contemporary Empirical Political Theory*, edited by Kristen R. Monroe, 73–86. Berkeley: University of California Press.

Hayek, Friedrich A. 1960. *The Constitution of Liberty*. Chicago: University of Chicago Press.

Hayek, Friedrich A. 1973. *Rules and Order*, Vol. 1 of *Law, Legislation and Liberty*. London: Routledge.

Hayek, Friedrich A. 1979. *The Political Order of a Free People*. Vol. 3 of *Law, Legislation and Liberty*. London: Routledge.

Jabłoński, Andrzej. 2012. Polityka. teoretyczna ewolucja pojęcia. In *Polityka i polityczność. Problemy teoretyczne i metodologiczne*, edited by Andrzej Czajkowski, and Leszek Sobkowiak, 11–42). Wrocław: Wydawnictwo Alta 2.

Karwat, Mirosław. 2009. Syndromatyczny charakter przedmiotu nauki o polityce. In *Demokratyczna Polska w globalizującym się świecie. I Ogólnopolski Kongres Politologii*, edited by Konstanty Adam Wojtaszczyk, and Andżelika Mirska, 175–188. Warszawa: Wydawnictwa Akademickie i Profesjonalne.

Karwat, Mirosław. 2010. Polityczność i upolitycznienie. Metodologiczne ramy analizy. *Studia Politologiczne* 17: 63–88.

Klementewicz, Tadeusz. 2010. *Rozumienie polityki: Zarys metodologii nauki o polityce.* Warszawa: Dom Wydawniczy Elipsa.

Klementewicz, Tadeusz. 2013. Politolog w labiryncie paradygmatów – pułapki eklektyzmu. In *Podejścia badawcze i metodologiczne w nauce o polityce*, edited by Barbara Krauz-Mozer, and Paweł Ścigaj, 31–43. Kraków: Księgarnia Akademicka.

Kukathas, Chandran. 1989. *Hayek and Modern Liberalism*. Oxford: Clarendon Press.

Laska, Artur. 2014. Polityka jako przedmiot badań w realiach instrumentalizacji politologicznego rozumu. *Przegląd Politologiczny* 4: 143–157. doi: 10.14746/pp.2014.19.4.10.

Long, Roderick T. 2006. Rule-following, Praxeology, and Anarchy. *New Perspectives on Political Economy* 2 (1): 36–46.

Long, Roderick. T., and Tibor R. Machan. 2008. *Anarchism/Minarchism: Is a Government Part of a Free Country?* London: Routledge.

Luhmann, Niklas. 1981. *Politische Theorie im Wohlfahrtsstaat*. München: Günter Olzog Verlag.

Muller, Jerry Z. 2007. The Limits of Spontaneous Order: Skeptical Reflections on a Hayekian Theme. In *Liberalism, Conservatism, and Hayek's Idea of Spontaneous Order*, edited by Louis Hunt, Peter McNamara, 197–209. New Your: Palgrave Macmillan.

Nozick, Robert. 1974. Anarchy, State, and Utopia, New York: Basic Books.

Pierzchalski, Filip. 2013. Polityka jako rozmyty przedmiot badań. In *Metafory polityki*, edited by Bohdan Kaczmarek, vol. 4, 35–51. Warszawa: Dom Wydawniczy Elipsa.

Popper, Karl. R. 2002. *The Open Universe: An Argument for Indeterminism*. London, New York: Routledge.

Posner, Richard A. 2005. Hayek, Law, and Cognition. *New York University Journal of Law and Liberty* 1 (0): 147–166.

Rhodes, R. A. W., Sarah A. Binder, and Bert A. Rockman. 2006. Preface. In *The Handbook of Political Institutions*, edited by R. A. W. Rhodes, Sarah A. Binder, and Bert A. Rockman, xii–xvii. Oxford: Oxford University Press.

Samuels, Warren J. 1999. Hayek from the Perspective of An Institutionalist Historian of Economic Thought: An Interpretative Essay. *Journal des Économistes et des Études Humaines* 9 (2/3): 279–290. doi: 10.1515/jeeh-1999-2-306.

Sandefur, Timothy. 2009. Some Problems with Spontaneous Order. *The Independent Review 14* (1): 5–25.

Servant, Régis. 2013. Spontaneous Emergence, Use of Reason and Constitutional Design: Is Hayek's Social Though Consistent? *Paper presented at the 14th Summer Institute for the History of Economic Thought, Richmond*, June 14–17.

Steele, G. R. 2002a. Hayek's Liberalism and Its Origins: His Idea of Spontaneous Order and the Scottish Enlightenment. Review of *Hayek's Liberalism and Its Origins: His Idea of Spontaneous Order and the Scottish Enlightenment,* by Christinia Petsoulas. *The Quarterly Journal of Austrian Economics* 5 (1): 93–95.

Ścigaj, Paweł. 2010. Granice w politologii jako wyznacznik tożamości dyscypliny. *Athenaeum* 26: 32–50.

Vanberg, Victor. 1994. Hayek's Legacy and the Future of Liberal Thought: Rational Liberalism versus Evolutionary Agnosticism. *Cato Journal* 14 (2): 179–199.

Wiśniewski, Jakub Bożydar. 2014a. Legal Polycentrism, the Circularity Problem, and the Regression Theorem of Institutional Development. *Quarterly Journal of Austrian Economics* 17 (4): 510–518.

10 The structure of human cognition

Hayek's concept of the cognitive system

Reflection on the theory of spontaneous order links with the issue of human cognitive limitations. It is the accessibility of knowledge that determines the human ability to deliberately form social structures and the distinction between spontaneous and planned order. And the very possibility of explaining social phenomena requires knowledge about these limitations. This means that the question of human cognition and the resulting problem of its limits are a key issue for the theory of spontaneous order. We may distinguish two essential aspects of this issue. The first is the structure and functioning of the cognitive system. The second is the issue of cognitive boundaries resulting from the existence of such a system.

The structure of the human cognitive apparatus

The concept of the cognitive system as formulated by Hayek (1952)[1] seems to be the right starting point for analysing the first issue. Hayek distinguishes two fundamental forms of order: physical and phenomenal (sensory). The first concerns processes taking place in the physical world. Within this order, Hayek distinguishes between the order of physical stimuli, which is external to the observer, and the neural order. The latter includes the system of neurons and synapses, along with the process of the creation and transmission of impulses. And the phenomenal order, also described as the sensory order, which encompasses all sensory qualities, formulates the sensations of the cognitive agent. This is done by classifying stimuli coming from the outside world. This classification functions thanks to the neural order that receives these stimuli. Within this order, individual cells react to a stimulus in a Boolean way, letting through or stopping an impulse. This means that the sensory order reacts to only some stimuli, while the occurrence of a specific reaction results from the sensitivity of the system to a given stimulus. This discrete way of reacting allows us to identify perception with classification of stimuli by the sensory order. It means that the relationship between these orders is isomorphic. However, this concerns the orders as a whole, and not the relationship between individual sensory experiences and the impulses of the nervous system. As Hayek wrote, "the sensory (or other mental) qualities are not in some manner originally attached to, or an original attribute of, the individual physiological impulses, but that the whole of these qualities is determined by the system of connexions by which the impulses can be transmitted from neuron to neuron" (Hayek 1952, 53).

This shape of the cognitive structures means that human perception of the world is determined by the classification apparatus. However, this apparatus is not static, with the response to the same situation (physical impact) always being identical. For while the neural order (and thus also the sensory order) is originally determined by the individual's genotype, it is subject to change in the process of his or her development. These changes result both from individual experiences acquired through the impact of external stimuli, and from the feedback mechanism modifying the classification apparatus in order to eliminate inconsistencies identified in it. Impulses affecting the neural order lead to changes in the structure of the network of neural connections, and thus to changes in the classification apparatus. However, the selective sensitivity of neurons means that not all stimuli are classified. This depends especially on the regularity of impulses, which by sensitising the order in a specific way modify it and enable the classification of incoming stimuli.

It is worth noting at this point that the approach describing cognition as a system of classification invokes the concept of category as proposed by Immanuel Kant (Horwitz 2000, 25). In both cases, reason, understood as the sensory order, is a system of classification, in which individual categories belong to the structure of the mind rather than the structure of the perceived world. However, unlike in Kant, for Hayek individual categories are not constant and immutable, but are subject to changes caused by individual experience.

Such a conception of the sensory order suggests that the structure of sensations is not a simple reflection of the physical properties of the external world, for there is no one-to-one mapping that would assign an individual impulse to an elementary sensory experience. So the phenomenal order, unlike the neural order, is not isomorphic with the external world order. Classification of an impulse that determines sensory impressions depends on past experiences that change the way this classification is made. And the shaping of the sensory order by these experiences makes the knowledge of individuals subjective. While the classification apparatus of individual people will be similar to each other because of the similarity of genotypes and environmental conditions, the impact of individual development makes them individual. This structure of the cognitive system forms the basis for the methodological subjectivism adopted by the spontaneous order theory.

The cognitive system as described by Hayek is intended for the survival of the individual. In this evolutionary approach, the system's task is to generate behaviours conducive to this goal in response to information coming from the environment. The stimuli must be classified by the sensory order as adequately as possible in relation to reality. This means that the structure of impulses within the classification apparatus should correspond to the structure of stimuli and thus to the structure of the external world. However, such adequacy is never absolute, due to a number of factors. Because of the discreteness and selectivity of the neuronal apparatus not all stimuli cause a corresponding change in the system of connections. Moreover, the classification apparatus is formed in a specific environment, which is local and so reproduces only part of the external world. The formation of cognitive structures is also influenced by the organism's internal environment and its anatomical preferences (Hayek 1952, 108–109).

The above approach means that the role of the classification apparatus is to provide a "map" reproducing relations and regularities occurring in the physical order. This map is a dynamic system of neural connections, shaped by past experience. At a given moment the impulses within this system create a model of the external world in which the person is currently located. This model represents the external environment and enables the interpretation of information that reaches the actor in the light of past experience. It also serves to anticipate possible consequences of both external phenomena and his or her own actions. The map, defining relations in the physical order, constitutes a specific set of possible phenomena, while the model indicates specific phenomena important for a given situation.

Boundaries of cognition

Hayek's reflections on the human cognitive apparatus can be seen as an extension of his earlier research in economics and the social sciences. Adopting an anti-naturalistic approach, Hayek points to the impossibility of explaining and predicting social phenomena using empirically testable laws, as in the natural sciences. The meanings that an individual attributes to their surrounding reality are subjective, so it is impossible to entirely reduce them to natural phenomena. This approach is used by the Austrian School of Economics to indicate the impossibility of aggregating the information needed for the central management of the economy, and forms the foundation of its liberal economic and political ideas.

However, this automatically raises questions about the anti-naturalistic assumption itself. Why should humans be incapable of explaining social phenomena in the same way as they do in the sphere of natural phenomena? Why should we assume that difficulties in formulating scientific laws concerning social reality require the use of fundamentally different research methods? Although social sciences do encounter predictive problems, this is not a sufficient justification for methodological dualism. Moreover, also in the natural sciences, the ability to predict future states of the observed phenomena is limited. This applies not only to quasi-deterministic systems described by quantum physics, but also to deterministic phenomena with a considerable degree of complexity. Although knowing the initial state makes it possible to predict future states of the system, the initial state is so complex that we cannot have complete knowledge of it. This leads to certain idealisations in the adopted model, producing a gap between predictions and the actual states of the examined phenomenon. Moreover, the ability to make predictions is constrained by finite computing powers. An example of this is the study of weather. The multitude of factors influencing atmospheric phenomena and the non-linearity of the processes shaping them makes it possible to formulate accurate forecasts only for a few days ahead.

The predictive limitations of the natural sciences, however, are relative, depending on the computational capacity and precision of description. In this sense, they are not permanent and impassable barriers to human cognition. The problem underlying them is solvable in principle. However, it is computationally intractable, which means that the amount of data needed to solve it is impossible to process due to limited temporal and computing resources (Marciszewski 2004, 7).[2] According to the proponents of methodological monism, the same concerns social phenomena, so the possibility of prediction within these sciences depends on the application of appropriate mathematical tools. As Klaus Mainzer (2007) points out, just like in the case of complex physical processes, models that take into account the non-linearity of the interdependence of factors shaping these phenomena are adequate for the description of social phenomena. According to Mainzer (2007, 123), the functioning of the mind can also be described by means of non-linear laws of complex system dynamics.

Hayek's idea of the human cognitive apparatus is an attempt to confront the problem of methodological dualism and possibly provide a justification for the anti-naturalistic position. It seeks an answer to the question of the limits of cognition, understood as a process of creating a model of reality. In other words, to what extent does this model, emerging in the course of impulse classification, make it possible to predict future states of the modelled reality? Using his idea of the mind, Hayek points out that in addition to the practical limitations resulting from the computational intractability of problems, human cognition also has an absolute limit, related to the existence of impassable and inherent barriers to cognitive processes. Hayek infers the existence of these barriers from the claim that the "apparatus of classification must possess a structure of a higher degree of complexity that is possessed by an object which it classifies" (Hayek 1952, 185). The degree of complexity should be understood as the smallest number of elements a structure must consist of in order to display all the attributes defining a given class of structures (Hayek 1967, 25). This means that for a given object to be classified, the degree of complexity is determined by the number of classes to which the object can be assigned.

The requirement for the classification structure to have a higher degree of complexity than the classified object follows from the fact that objects belonging to various classes differ from each other by at least one attribute, against which the classification is made. This means that the number of object classes created in the classification is the number of possible combinations of attribute classes to which these objects can be assigned. In the context of perceiving the mind as a classification apparatus, this produces cognitive constraints. In particular, it indicates that it is impossible to "fully" explain the mind itself.[3] This would require a cognitive apparatus that would be more complex than the mind itself. Even if we assume that such an apparatus exists and is available, this explanation would also require an explanation of the apparatus itself. This, in turn, would require the use of a cognitive device that would be even more complex than the apparatus. Consequently, trying to explain the mind leads to an infinite process of selecting more and more complex apparatuses.

Of course, a lack of complete cognition is not equivalent to a lack of access to any knowledge about complex structures. Our knowledge about the phenomena we examine is gradable, because classification of impulses by the cognitive apparatus varies in the degree of detail.[4] The greater the complexity of a given structure, the less detailed is the explanation and, consequently, the less able it is to predict future states.

Such a cognitive constraint is absolute because unlike practical constraints resulting from computational intractability it does not depend on the available time or the computational capacity of the mind. It is also enmeshed in the problem of self-reference that occurs when cognition is referenced to cognition itself. This indicates that the mind can never understand and explain itself.

Noting the issue of the complexity of the cognitive apparatus and the corresponding absolute cognitive limitations has a bearing on the validity of the position of methodological dualism. As Gorazda (2013, 101–102) notes, the importance of the relationship between the degree of detail of an explanation and the degree of complexity of the studied phenomenon makes the former dependent on the degree to which the phenomenon depends on the operation of the cognitive apparatus itself. The greater the impact of the human mind on the examined object, the lower the degree of detail of the possible explanation.

> The more the object moves away from our cognitive apparatus, the more detailed and perfect the model can be. If the object of our explanation approaches the area of our cognitive apparatus, the model will be less detailed. [...] If the object of cognition is the dynamics of movement of a rigid body, this dependence is expressed solely in the way this movement is perceived. The movement of the body itself is independent of the apparatus. However, if the object of cognition is actions taken or decisions made by individuals or communities, this dependence is very strong.
>
> (Gorazda 2013, 101–102)

Hence, the nature of the explanation differs from discipline to discipline. In the social sciences, the object of research is concentrated around the human being and society, which produces a greater degree of dependence on human cognition than is the case in the natural sciences. Consequently, an explanation formulated within the social sciences is characterised by a greater degree of generality. This finds a natural reflection in the problem of making predictions in the social sciences, since prediction requires detailed knowledge of a given phenomenon and access to this knowledge is limited in this field. It also undermines the ability to falsify the formulated theories, especially by revealing the need to weaken the Popperian falsification condition as a determinant of being scientific (see Hayek 1967, 22–42).[5]

The embodied-embedded mind paradigm

The presented concept of the apparatus of the mind, understood as a sensory order, not only makes it possible to formulate an argument in favour of an anti-naturalistic approach, but is also a good starting point for the reflection on the problem situation, that is the issue of human cognitive boundaries. These limits determine the actions taken by agents, which indicates the importance of cognitive science in the study of social phenomena.

Cognitive science deals with the issue of the mind and cognitive processes, drawing strongly on many disciplines such as philosophy, anthropology, linguistics, neurobiology and psychology. In its initial period of development (from the 1950s to the 1980s), it was dominated by the approach of computational functionalism (Hohol 2011, 151). Functionalism considers mental states in a given cognitive system as functional states. This means that a given mental state is described by the states which create it and which it causes. Since the brain is an information-processing system, mental states are correlations between brain input and output signals. This makes the conscious mind (as consisting of mental states) a function performed by the brain to process the information it receives.[6] Computational functionalism makes the additional assumption that this function is computable in Turing's sense. This means that it is a mathematical function which can be described as an effective mechanical operation with a finite number of steps. This makes it an intuitive equivalent of an algorithm. The machine carrying out this process is referred to as a Turing machine. It is a kind of model of an "ideal computer", which implements algorithms without making mistakes.[7]

This approach, also called symbolism or computationalism, means that thinking is a computational process, consisting of transforming symbols. And the concept of intelligence is revealed in the ability to solve problems through the process of transforming symbols that represent a given problem. This leads to the perception of what is commonly referred to as the mind, as an algorithm (a computer programme) performed by a computer, the brain. Implementation of these algorithms defines thought processes such as learning or problem solving (Anderson 1987; Marr 2010). Consequently, for computationalism, it is the processes at the algorithmic level that determine human behaviour.

In later years, the symbolic approach was largely replaced by approaches saying that cognitive processes do not have to be algorithmic, but can also be algorithmically approximable (Hohol 2013, 90). The approach described as "connectionism" plays a particularly important role. It models the cognitive apparatus not as a symbol-processing Turing machine, but as using artificial neural networks. This idea was born out of a desire to map certain aspects of the human nervous system. Certain similarities between the above approaches make it possible to place all of them within a single research paradigm, called the computer paradigm (Hohol 2013, 88–104). In particular, it makes a distinction between the implementation and representation levels. The former is the level of symbol processing, the latter consists of the biological mechanisms in which processes from the first level are implemented.

The second generation of cognitive science rejects this understanding of the cognitive apparatus. Its embodied mind approach highlights the importance of the human body in the cognitive process. This process is a feedback loop between the mind and the body "housing" it. While the cognitive system affects the human body, the body also shapes the mind and cognition through what it experiences in contact with its environment. This means that the ability to reason is dependent on our carnality. As Lakoff and Johnson write:

> Reason is not disembodied, as the tradition has largely held, but arises from the nature of our brains, bodies, and bodily experience. This is not just the innocuous and obvious claim that we need a body to reason; rather, it is the striking claim that the very structure of reason itself comes from the details of our embodiment. Thus, to understand reason we must understand the details of our visual system, our motor system, and the general mechanism of neural binding.
>
> (1999, 4)

The emergence of the embodied mind approach is associated with changes in linguistics and the arrival of cognitive linguistics.[8] Cognitive linguistics holds that the emergence and functioning of language involves mental, perceptual and motoric processes and that meaning is created through the interaction between the human body and the surrounding external world.

An important role is played here by the concept of metaphor, which is not so much a means of linguistic expression as a cognitive mechanism for understanding a given issue through the use of a conceptual pattern from another issue, more familiar or described in more detail, and assigning its structural features to the original one (see Lakoff and Johnson 2003). This allows for the transfer of meanings from one structure to another, and enables the creation of abstract concepts, based on specific concepts produced by the embodied mind in the course of its interaction with the world.

Cognitive linguistics breaks with the traditional understanding of language, blurring the difference between the syntactic rules and meanings (semantics) of a language. A symbol is a reference to the sphere of human experience and mental states, that is not to the external world, but to its interpretation by the embodied mind. This means that language is not reduced to a meaning-independent syntax. This leads to the rejection of Noam Chomsky's generativism, which proclaims the existence of universal grammar common to all languages (see Pinker 1995). According to supporters of generativism, grammar emerged from genetic conditions resulting from the phylogeny of the human species. In contrast, according to the concept of the embodied mind, the emergence of grammatical structures is a cultural and historical phenomenon (Tomasello 2003, 9). The genetic foundation is common to both syntax and semantics, which rules out their clear-cut differentiation.[9] Unlike in functionalism, language is not a product of abstract symbolic structures, but of bodily interactions of the agent. This leads to rejection of the concept of humans as acting on the basis of meaningless symbols as formulated by the computer paradigms:

> There is no such thing as a computational person [...] whose mind somehow derives meaning from taking meaningless symbols as input, manipulating them by rule, and giving meaningless symbols as output. Real people have embodied minds whose conceptual systems arise from, are shaped by, and are given meaning through living human bodies. The neural structures of our brains produce conceptual systems and linguistic structures that cannot be adequately accounted for by formal systems that only manipulate symbols.
>
> (Lakoff and Johnson 1999, 6)

The importance of the environment in cognitive processes concerns not only the natural world but also the social environment. The mind is not only embodied but also embedded in culture[10] and social interactions (embedded mind).[11] The importance of cultural embedding is important because of the problem encountered in attempts to explain the uniqueness of human cognition based on biological evolution. As Michael Tomasello (1999, 2) notes:

> The 6 million years that separates human beings from other great apes is a very short time evolutionarily, with modern humans and chimpanzees sharing something on the order of 99 percent of their genetic material – the same degree of relatedness as that of other sister genera such as lions and tigers, horses and zebras, and rats and mice. Our problem is thus one of time. The fact is, there simply has not been enough time for normal processes of biological evolution involving genetic variation and natural selection to have created, one by one, each of the cognitive skills necessary for modern humans to invent and maintain complex tool-use industries and technologies, complex forms of symbolic communication and representation, and complex social organizations and institutions.

The uniqueness of the human in his cultural evolution cannot be attributed only to the ability to imitate and attribute intentions to other agents. Early humans' form of cooperative communication allowed the coordination of perspectives on the external situation, thus introducing cooperation into individual intentionality. This transformed individual intentionality into joint intentionality embracing new forms of cognitive representation (Tomasello 2014, 33). The human species is characterised by a unique faculty of social cognition, which contributed to the emergence of culturally conditioned, purposeful learning processes. Social cognition produced the ratchet effect, allowing for the social transmission and accumulation of knowledge. It prevents the loss of knowledge once created and a continuous retreat to previous stages of cultural evolution. Since man is already born in a specific cultural context, subsequent generations do not have to produce it anew, but acquire it in the process of learning, as well as producing new patterns within it. And the very idea of the cumulative process of cultural evolution is compatible with the theory of spontaneous order as recognising the social nature of man and indicating the limitations of conscious and planned formation of social reality.

A similar approach can be found in Hayek's concept of the dynamic mind. As in the paradigm of the embodied mind, the sensory order, although originally shaped by the genotype, is subject to change in the course of the person's interaction with the world. This applies not only to the physical world but also to the social world. As Hayek wrote:

> What we call mind is not something that the individual is born with, as he is born with his brain, or something that the brain produces, but something that his genetic equipment (e.g., a brain of certain size and structure) helps him to acquire, as he grows up, from his family and adult fellows by absorbing the results of a tradition that is not genetically transmitted.
>
> (Hayek 1988, 22)

It is therefore not a static structure, independent of the outside world, but has a dynamic ("semipermanent") character. This means that, as in the paradigm of the embodied and embedded mind, the human body co-shapes the mind and cognition through what it experiences in contact with its environment.

The Hayekian concept of mind adopted by the spontaneous order theory is also a rejection of the evolutionary psychology approach.[12] Evolutionary psychology examines human behaviour with the use of the evolutionary approach, largely continuing the tradition of sociobiology. From the point of view of spontaneous order theory, this approach is problematic as it is focused on biological evolution and ignores the issue of individual self-awareness as the causative factor. Consequently, evolutionary psychology reduces man to a thoughtless automaton. This in turn leads to ignoring issues following from the fact that man is not only an object, but also the subject of cognition. In particular, this leads to downplaying or ignoring the role of culture in shaping the human mind and behaviour. Meanwhile, the human ability to think abstractly and the role of cultural factors in decision-making can lead to behaviours contrary to those driven by mental dispositions produced by natural selection.

Notes

1 In many aspects, Hayek can be considered a pioneer of current concepts in the philosophy of the mind and cognitive sciences. In particular, the concept of

embodiment, so close to contemporary approaches to the human mind, can be seen in his analysis. Nevertheless, his work in this area has remained largely unnoticed (see Steele 2002b; Gorazda 2012; Zonik 2013).

2 Strictly speaking, the problem is computationally intractable when there is no algorithm solving this problem in time bounded by polynomial function of size of input data. An illustration of the problem of computational complexity is provided by the travelling salesman problem. Given a set of n cities and distances between each pair of cities the travelling salesman problem asks what is the shortest route encompassing all cities and ending at the starting point (see Marciszewski 2004, 13–14). For a task so defined, the number of routes to be considered is equal to the number of orders on a n-element set and amounts to $n!$ This means that the number of routes increases exponentially with the number of cities. If we have four cities, it means choosing from $4! = 24$ routes, but with ten cities it is necessary to compare $10! = 3{,}628{,}800$ routes. For $n = 24$ the number of possible configurations is so large that, assuming the capacity to check one million routes per second, the time it would take to check all the possibilities is longer than the age of the universe.

3 The concept of "full" or complete explanation means that there are no limits as to the degree of detail in explaining the examined phenomenon.

4 The concept of the degree of detail may be demonstrated with the erotetic model of explanation (see Gorazda 2013). In this model, the explanation is defined by a given problem and aims to answer the question posed by this problem. This means that both the question and the answer depend on the existing knowledge. In this sense it is not possible to create a complete model, which would give a "full" explanation of the investigated phenomenon. However, it should be remembered that this model concerns scientific explanation, which is intersubjective, while Hayek focuses on explanation within a single mind. Having said that, the degree of detail of an explanation can be interpreted as the number of correctly formulated problem questions which this explanation answers.

5 Some elements of this approach can be found in Popper himself. In his work *The Myth of the Framework* (Popper 1994) he indicates that the social sciences use the explanation "in principle", not "in detail". This suggests that the significance of scientific openness, which Popper opted for, is not constrained to openness to falsification, but can be understood more generally as openness to various research methods.

6 In particular, this implies that consciousness can be attributed to any structure capable of adequate information processing (see Klinowski 2008, 37).

7 In this sense, the digital computer (equipped with operating systems) is an implementation of Turing machine.

8 A discussion of linguistic concepts of cognition can be found in, among others, Gemel (2015).

9 A critique of Chomsky's concept of generative grammar can be found in, among others, Brożek (2016, 102–124).

10 Culture is understood here as "the total pattern of human behaviour and its products embodied in thought, speech, action, and artifacts [technologies] and dependent upon man's capacity for learning and transmitting knowledge" (Gove 1971, 552).

11 So instead of the embodied mind paradigm, we can speak of the embodied-embedded mind paradigm.

12 A discussion of the issue of applying evolutionary psychology as a research perspective in political science can be found in Ścigaj (2014).

Bibliography

Anderson, John R. 1987. Methodologies for studying human knowledge. *Behavioral and Brain Sciences* 10 (3): 467–477. doi: 10.1017/S0140525X00023554.

Brożek, Bartosz. 2016. *Granice Interpretacji*. Kraków: Copernicus Center Press.

Gemel, Aleksander. 2015. *Językowy model poznania: kognitywne komponenty w kontynentalnej filozofii języka*. Łódź: Wydawnictwo Uniwersytetu Łódzkiego.

Gorazda, Marcin. 2012. Normativity According to Hayek. In *The Many Faces of Normativity*, edited by Jerzy Stelmach, Bartosz Brożek, and Mateusz Hohol, 223–256. Kraków: Copernicus Center Press.

Gorazda, Marcin. 2013. Granice wyjaśnienia naukowego. Część II. *Zagadnienia Filozoficzne w Nauce* 52: 53–106.

Gove, Philip B., ed. 1971. *Webster's Third New International Dictionary of the English Language Unabridged*. Springfield: Merriam-Webster.

Hayek, Friedrich A. 1952. *The Sensory Order. An Inquiry into the Foundations of Theoretical Psychology*. London: Routledge and Kegan Paul.

Hayek, Friedrich A. 1967. *Studies in Philosophy, Politics and Economics*. London: Routledge and Kegan Paul.

Hayek, Friedrich A. 1988. *The Fatal Conceit: The Errors of Socialism*. London: Routledge.

Hohol, Mateusz. 2010. Umysł: system sprzeczny, ale nie trywialny. *Zagadnienia Filozoficzne w Nauce* 47: 89–108.

Hohol, Mateusz. 2013. *Wyjaśnić umysł: struktura teorii neurokognitywnch*. Kraków: Copernicus Center Press.

Horwitz, Steven. 2000. From the Sensory Order to the Liberal Order: Hayek's Non-rationalist Liberalism. *The Review of Austrian Economics* 13 (1): 23–40. doi: 10.1023/A:1007850028840.

Klinowski, Mateusz. 2008. Funkcjonalizm obliczeniowy – kilka uwag z perspektywy ewolucyjnej. *Rocznik Kognitywistyczny* 2, 37–43.

Lakoff, George, and Mark Johnson. 1999. *Philosophy in the Flesh: The Embodied Mind and Its Challenge to Western Thought*. New York: Basic Books.

Lakoff George, and Mark Johnson. 2003. *Metaphors We Live By*. Chicago: The University of Chicago Press.

Mainzer, Klaus. 2007. *Thinking in Complexity: The Computational Dynamics of Matter, Mind, and Mankind*. 5th ed. Berlin: Springer-Verlag.

Marciszewski, Witold. 2004. Nierozstrzygalność i algorytmiczna niedostępność w naukach społecznych. *Filozofia Nauki* 12 (3–4): 3–51.

Marr, David. 2010. *Vision: A Computational Investigation into the Human Representation and Processing of Visual Information*. Cambridge, London: The MIT Press.

Pinker, Steven. 1995. *The Language Instinct. The New Science of Language and Mind*. New York: Harper.

Popper, Karl. R. 1994. *The Myth of the Framework: In Defence of Science and Rationality*. London: Routledge.

Steele, G. R. 2002b. Hayek's *Sensory Order. Theory and Psychology*, 12(3), s. 125–147.

Ścigaj, Paweł. 2014. Podejście ewolucyjne. Nowa perspektywa w badaniach politologicznych. In *Odmiany współczesnej nauki o polityce*, Piotr Borowiec, Robert Kłosowicz, and Paweł Ścigaj, vol. 1, 219–249. Kraków: Wydawnictwo Uniwersytetu Jagiellońskiego.

Tomasello, Michael. 1999. *The Cultural Origins of Human Cognition*. Cambridge, MA: Harvard University Press.

Tomasello, Michael. 2003. *Constructing a Language. A Usage-Based Theory of Language Acquisition*. Cambridge: Harvard University Press.

Tomasello, Michael. 2014. *A Natural History of Human Thinking*. Cambridge, MA: Harvard University Press.

Zonik, Przemysław. 2013. Neuroantropologiczne ujęcie związków pomiędzy mózgiem, działaniem i kulturą na przykładzie percepcji. In *Umysł i Kultura*, edited by Urszula Maja Krzyżanowska, Joanna Zonik, and Przemysław Zonik, 95–131. Lublin: Werset.

Conclusion

The theory of spontaneous order can be described as a research position in which social phenomena are seen in the context of the mutual coordination of human activities. It says that action of itself produces a state of uncertainty, resulting from the absence of full knowledge about the effects of that action. It also emphasises the creative nature of the agents' conduct. This makes it possible to distinguish two fundamental issues raised by this theory.

The first issue concerns limitations in the planned shaping of social order. Since knowledge determines the actions of individuals, and thus the social phenomena resulting from these actions, this theory denies the omnipotence of man in the design and purposeful shaping of social structures. The planned coordination of all human activities requires collecting all necessary knowledge within a central decision-making body. However, from the point of view of the spontaneous order theory, the dispersion and incompleteness of the knowledge available to man are inherent aspects of human action. This raises the question of the limits of the planned shaping of social order, and indirectly about the limits of human cognition.

The second issue concerns the phenomenon of spontaneous order itself. The absence of full knowledge means that the order formed through processes of coordination of activities does not have to be the intended goal of these activities. This perspective leads the spontaneous order theory to reject the approach according to which institutions forming society are or should be the product of a deliberate plan. This offers us the possibility to grasp and explain the emergence and functioning of a number of institutions and social phenomena that are not the result of individual decision-making or collective consensus aimed at their establishment. In particular, the spontaneous order theory indicates that many institutions with a complex structure which might seem to be a deliberate product of a plan of some omniscient decision-making body result from the bottom-up, unintentional actions of individuals.

As the theory is not constrained to specific forms of human activity, it potentially offers a broad field of application also in the area of political science. It should be noted as well that political science focuses on studying the active and planned shaping of social order, expressed in decisions and actions aimed at changing social structures. Consequently, the spontaneous order theory does not seem to have any significant application within political science as it deals with explaining unplanned structures. This concerns especially the narrow understanding of politics focused on state institutions. Much more important is the application of this theory in the context of the limits of planned social order formation. Paradoxically, it is the teleological aspect of political phenomena that causes the problem of human cognitive limitations indicated by the spontaneous order theory to resonate. As a result, the theory allows for attempts to explain the dysfunctionality of social and political structures by invoking the issue of ignorance and uncertainty. It also allows us to recognise and comprehend the coordinating aspect of dynamic and unplanned social processes.

This theory seems particularly useful in relation to issues raised by the theory of public choice. As the theory of public choice is based on neoclassical economics, criticism of this economics by the Austrian School can be transferred to political science. The theory of spontaneous order can also formulate solutions to problems that arise from public choice theory. In particular, implementation of the spontaneous order theory leads to the rejection of the ideas of rational ignorance and rational irrationality as defining the mechanisms of choice. Instead, the theory uses the concept of radical ignorance, which has its origin in the primacy of the problem of knowledge, creating an alternative approach that does not require the assumption of omniscience or irrationality of individuals. In the context of human creativity, this leads to the rejection of the concept of market and state failure in the form presented by the theory of public choice.

The particular importance of the problem of knowledge also produces the possibility of applying the spontaneous order theory to the question of the effectiveness of mono- and polycentric systems of coordinating human actions. Unlike public choice theory and the Bloomington School, it points out that the crucial issue in the problem of coordination is not the number of agents in the system of interaction, but the availability of information. It also sees the issue of embedding the source and carrier of information in the human being and the importance of character and complexity of the subject of decision. This leads to criticism of the description of democratic systems using models based on market competition mechanisms. The focus on the radical ignorance of man as the basis for distinguishing between these structures also leads to emphasising the role of the beliefs and ideas used to interpret reality. The fact that political decisions are made in conditions of uncertainty means that they are largely based on social heuristics and the extrapolation of moral beliefs.

Another area of possible use of the spontaneous order theory is the problem of fragile states. Since it emphasises the importance of informal social and political institutions and notes the dynamic nature of social processes, it can be applied to the analysis of institutional changes in regions without a stable political and legal system. Rejecting the state-centric approach, the spontaneous order theory treats the problem of state dysfunctionality as relating to the whole spectrum of social phenomena associated with the instability of social, political, legal and economic structures. Discarding the perception of the state as an institution necessary for the existence of social and legal order, it treats it as a factor that interferes with other institutions operating in a given area. The state is therefore just one of many institutions shaping this order. The issue of state failure makes the spontaneous order theory question the approach in which factors determining state dysfunctionality are identified with processes that weaken state-forming processes. It rejects not only attempts to assess and model fragile states on idealised conceptions of statehood or relatively efficient Western states. It also questions constraining analyses to the issue of adapting new state structures to local cultural conditions, since this preserves the assumption that a centralised decision-making system is necessary. The issue of state dysfunctionality should not be reduced to the effectiveness of decision-making mechanisms within a centralised system of power, but should be based on a broader question of the legitimacy of consolidating decision-making processes. It cannot be ruled out that the absence of a state may serve cooperation and social order better than the existence of inefficient state structures weakening local institutions.

The existence of cognitive human limitations and the possibility of the emergence of a spontaneous order are also applied to the concept of the veil of ignorance. The spontaneous order theory points to the problem of the radical ignorance of decision-makers for this idea, and thus to the possible discrepancies between planned and actual effects of decisions. Seeing the risk of the pretence to knowledge that emerges from the planned shaping of order, it emphasises the importance of learning through imitation and the transmission of information in the processes of formation of social institutions.

The spontaneous order theory seems to provide a valuable approach to analysing the formation and change of social structures. However, it is necessary to point out the doubts that can be raised by the dichotomous distinction between two forms of order – spontaneous and planned – as applied to the complex reality of social phenomena. The solution here is to invoke the Popperian concept emphasising the importance of the context of the research problem and the subservient role of the terms applied in identifying problem situations and trying to solve them.

The fundamental problem that arises from the research perspective adopted by the spontaneous order theory is not the ambiguity of the terms employed, but the issue of human cognitive limits. This highlights the issue of cognitive processes as crucial for the development of this theory. The starting point here is Hayek's concept of the apparatus of the mind, understood as a sensory order. As in the paradigm of the embodied mind, this order is dynamic and undergoes transformations in the course of the actor's interaction with the world. Such a concept of the mind also leads the theory of spontaneous order to reject the approach of evolutionary psychology as ignoring the issue of individual self-awareness and the role of man as an object of cognition.

Index

Printed in the United States
By Bookmasters